Pollution and Climate Change

Editors

Dr. Mamta Sharma Dr. Hukam Singh

Dr. Upendra Singh

Pustak Bharati
Toronto Canada

Editors : Dr. Mamta Sharma
Dr. Hukam Singh
Dr. Upendra Singh

Book Title : Pollution and climate change

Cover Picture : By Dr. Anil Kumar Chhangani, D.Sc

Published by :
Pustak Bharati (Books India)
180 Torresdale Ave, Toronto Canada M2R3E4
email : pustak.bharati.canada@gmail.com
Web : www.pustak-bharati-canada.com

Published for
Raj Rishi Government Autonomous College,
Alwar, Rajasthan, India

Financial Assistance
Rashtriya Uchchatar Shiksha Abhiyan
(RUSA-2.0)

Copyright ©2023

ISBN : 978-1-989416-44-0

© All rights reserved. No part of this book may be copied, reproduced or utilised in any manner or by any means, computerised, e-mail, scanning, photocopying or by recording in any information storage and retrieval system, without the permission in writing from the editors.

Preface

"Our planet is slowly dying, and if we don't do anything about it soon enough, it would eventually begin to deteriorate and everything would be used. The world would become a barren place without any resources. We need to cater to the needs of our planet, and we need to change our life styles so that it becomes beneficial to the planet. We need to become much more eco-friendly, so that no harm is dealt to the planet by our existence. Many people don't realize that they waste large amounts of energy and other resources in various unnecessary things that could otherwise be saved."

This series of books is an extension of the 3 days international conference on **Multidisciplinary Approach Towards Sustainable Develop-ment and Climate Change for A Viable Future (ICMSDC-2022)** held from 12th-14th August 2022 at Raj Rishi Government Autonomous College, Alwar, Rajasthan.

We are very happy and delighted to publish our series of books which are accumulation of research papers of knowledgeable experts in the field of sustainable development and climate change.

Climate change is the most significant challenge to achieving sustainable development, and it threatens to drag millions of people into grinding poverty. At the same time, we have never had better know-how and solutions available to avert the crisis and create opportunities for a better life for people all over the world. Climate change is not just a long-term issue. It is happening today, and it entails uncertainties for policy makers trying to shape the future.

There is a dual relationship between sustainable development and climate change. On the one hand, climate change influences key natural and human living conditions and thereby also the basis for social and economic development, while on the other hand, society's priorities on sustainable development influence both the greenhouse gas emissions that are causing climate change and the vulnerability.

Climate policies can be more effective when consistently embedded within broader strategies designed to make national and regional development paths more sustainable. This occurs because the impact of climate variability and change, climate policy responses, and associated socio-economic development will affect the ability of countries to achieve sustainable development goals. Conversely, the pursuit of those goals will in turn affect the opportunities for, and success of, climate policies.

With these books, we aim to reach to as many people as we can, and spread awareness about sustainable development and climate change and its in-depth analysis through our didactic research papers. We hope that the thought with which ICMSDC-2022 was executed is taken forward through this series of books and the inception of an idea of saving the environment is rooted in the minds of our readers. The articles in these books have been contributed by eminent research scholars, scientists, academicians and industry experts whose contributions have enriched this book series. We thank our publisher, Pustak Bharati, Toronto, Canada for joining us in this initiative and helped in publishing this series of books.

Finally, we will always remain indebted to all our well-wishers for their blessings, without which ICMSDC-2022 and series of these book would have not come into existence.

Financial Assistance provided by Rashtriya Uchchatar Shiksha Abhiyan (RUSA-2.0) is gratefully acknowledged.

Dr. Mamta Sharma
Dr. Hukam Singh
Dr. Upendra Singh

Contents

   Preface

1. Reduce, Reuse and Recycle is very Important to save The Planet ... 1
   Dr. Mamta Sharma
   Dr. Hukam Singh
   Dr. Upendra Singh

2. A Systematic Review on Prevalence of Dental Fluorosis in Rajasthan ... 12
   Dr. Neha Goyal,
   Dr. Neetu Agrawal,
   Dr. Shiv Singh Dulawat

3. Climate Change and its Effect on Environment ... 26
   Nirmala Bansal

4. Novel 1,3,4-oxadiazole Derivatives : Design, Characterization and Biological evaluation ... 31
   Jyoti Sharma, Nidhi Agarwal

5. Nature's Insurgence and Solidarity : Milton's Travel to Corporal and Supernatural Environment in *Paradise Lost* ... 45
   Dr. Nipun Chaudhary

6. Mycoremediation : A Sustainable Clean Technology for Better Tomorrow ... 51
   Dr. Geetanjli, Simranjot, Kiranjeet Kaur

7. Multidisciplinary Approch Towards Sustainable Development and Climate Change ... 73
   Prof. (Dr.) Rakesh Daiya

8. पर्यावरण संरक्षण के कानूनी पहलू ... 84
   हरकेश मीणा

9. "Seed to Seedling Transmission and Phytopathological Effects of *Pseudomonas Aeruginosa* Causing Brown Soft Rot Disease in Onion" ... 90
   Laxmi Meena[1] and Laxmikant Sharma[2]

10.	Spectrophotometric Determination of Labetalol Using CDNB Reagent B. Eswara Naik, C. N. Rao C. Narasimha Rao	95
11.	सतत आर्थिक विकास व पर्यावरण संरक्षण : अन्तर्निर्भरता तथा सामाजार्थिक लागतें श्रीमती कविता शर्मा	104
12.	Integration of Logistic Innovative Approach for Solving Multi-Objective Transportation Problem Jitendra Singh and Anju*	117
13.	Impact of War on Biodiversity with particular reference to Critically Endangered Species (An attempt to find solution to end up wars) Prof. Krishan Kumar Sharma	128
14.	Impact of Climate Change on Indian Agriculture Bharat Yadav Dr Manju Yadav	131
15.	Indoor Air Quality : A Way to Sustainable Buildings Dr. Bhawana Asnani	145
16.	Physico-Chemical Characterization of Farmland Soil in Rajuri Village of Rahata Taluka, Ahmednagar District, Maharashtra (India) Vikhe A.S, Vikhe P.S, Kharde H.S	153
17.	House Plants Combating Indoor Pollution and Relieving Sick Building Syndrome Madhulika Parmar	163
18.	Duration of Parental Care in Blackbucks in Captivity Sonia Yadav[1] and Dr. Raksha Modi[2]	170
19.	Impact and use of Fertilizers in Indian Agriculture Ms. Komal Bansal	181
20	Sustainable Development : A New Approach towards Mitigation of the Conflict between Development and Environment through the Lens of Indian Judiciary Dr. Suruchi Saurabh	191

1. Reduce, Reuse and Recycle is very Important to save The Planet

Dr. Mamta Sharma*
Dr. Hukam Singh**
Dr. Upendra Singh ***

Primer :

The new decade is upon us, and many of us have the same goal in mind: to be more eco-friendly. Living a more sustainable lifestyle can be very rewarding – but also a bit daunting. Thankfully there are four easy-to-follow stepping stones that we can apply to everyday life: reduce, reuse, recycle and repurpose.

Reduce

The first step to this eco-friendly quartet is to reduce, which seems self-explanatory – just use less. Each year, we use 1 billion plastic shopping bags, creating 300,000 tons of landfill waste. You can start reducing your use of plastic by:

- Bringing reusable bags to the grocery store and buying in bulk whenever you can,
- Using a washcloth instead of a paper towel when cleaning up kitchen messes,
- Taking shorter showers,
- Hand-washing dishes instead of using the dishwasher,
- And avoiding single-use plastics (plates, bowls, cups, Styrofoam, straws, water bottles, etc.)
  . Whether you're reducing your waste, your water usage or your buying habits, a little effort goes a long way for the environment.

Reuse

Next up to bat is reuse, which is as simple as it sounds. When you reuse products, you're able to make use of them multiple times or in multiple ways. In less than 15 years, worldwide waste is expected to double.2 So, instead of throwing away, find a way to reuse items you would otherwise toss! Some super simple ways to reuse what you already have and keep trash out of landfills are:

- Purchase a reusable water bottle instead of using single-use plastic bottles.
- Use your old toothbrush to scrub small, hard-to-reach areas when cleaning the house.
- Turn your empty coffee tin into a storage container for your household items.
- Pop your used tea bags in the fridge then place them on your eyes for a natural eye de-puffer.
- Turn a broken frame into an earring holder by attaching wire across the frame and hanging earrings from a wire.
- Cut the top off of an old soda bottle and hang it from a tree with birdseed for an easy bird feeder.

There are endless ways to reuse items that would typically be going to waste, but the best way you will be successful with this step is to be CREATIVE!

Recycle

The third "R" on our list, recycling, is the process of turning waste into new materials or objects. In some cases, producing waste is inevitable. However, it is important to be aware of the waste you are producing and how to recycle as much of that waste as possible. For example:

If you have to use a plastic spoon, clean it thoroughly and make sure it is recycled.

Find a local recycling center to recycle old electronics or furniture.

If you can't find a use for old plastic or glass bottles and jars, give them a rinse and make sure they end up in the recycling bin.

The recyclable materials in the U.S. waste stream would generate over $7 billion if they were recycled. Of course, it is important to produce as little waste as possible, but when you do produce waste, determine how much of it can be recycled, clean it off and throw it in the recycling bin.

Repurpose

Perhaps the most creative of the four R's is to repurpose by taking a product you don't use or haven't used in a while and giving it a new life somewhere else. Here are a few of our favourite ways to repurpose items commonly thrown away that could be used for another job:

Have an old t-shirt? Instead of throwing it away, cut it up into squares or rectangles and BOOM, you've got yourself new wash rags.

Have an empty milk carton? Poke holes in the top of the cap and you've got yourself a DIY watering can.

If you have any old jewellery, remove it from its base and hot glue a magnet to the back for new and improved fridge magnets.

Reinvent your old CDs into cool coasters for your coffee table.

There are so many ways to repurpose items around your house, it just takes a little bit of creativity (and a little help from the internet) to perfect any DIY repurposing project.

If you start implementing the four R's into your everyday life, the difference you will have in the environment will become exponential. Once you've perfected them, start sharing them with your friends and showing them all the ways, they can easily reduce, reuse, recycle and repurpose. Who knows, maybe you'll pick up some new tips along the way!

If you have heard of something called the "waste hierarchy," then you may be wondering what that means. It is the order of priority of actions to be taken to reduce the amount of waste generated and to improve overall waste management processes and programs. The waste hierarchy consists of 3 R's as follows:

Reduce

Reuse

Recycle

Commonly called the "three R's" of waste management, this waste hierarchy is the guidance suggested for creating a sustainable life. You might be wondering how you can incorporate these principles into your daily life.

They are not that hard to implement. All you need is to bring a small change in your daily lifestyle to reduce waste so that less amount of it goes to the landfill which can reduce your carbon footprint.

"It makes a big difference to recycle. It makes a big difference to use recycled products. It makes a big difference to reuse things, to not use the paper cup – and each time you do, that's a victory"

~Emily Deschanel

"The three R's – reduce, reuse and recycle– all help to cut down on the amount of waste we throw away. They conserve natural resources, landfill space and energy. Plus, the three R's save land and money that communities must use to dispose of waste in landfills. Siting a new landfill has become difficult and more expensive due to environmental regulations and public opposition. "By refusing to buy items that you don't need, reusing items more than once and disposing of the items that are no longer in use at appropriate recycling centers, you can contribute towards a healthier planet.

The First 'R'– Reduce

The concept of reducing what is produced and what is consumed is essential to the waste hierarchy. The logic behind it is simple to understand – if there is less waste, then there is less to recycle or reuse.

The process of 'reduce' begins with an examination of what you are using, and what it is used for. There are three simple steps to assessing the reduction value of an item or process –

Step 1 : Is there something else that can be used for this purpose?

Using multi-use items is essential to beginning reduction. One example would be a coffeepot and a cappuccino maker. Both of them distinctly do different things, but you can buy a coffeepot that has a steaming attachment on it so it can do both.

The purchase of one item means that you don't use two. It reduces the amount of production and the amount of waste packaging material that will be generated.

Step 2 : Is this something that needs to be done?

A lot of our waste material comes from items that are considered to be "disposable." Not in the sense that you use something once and then throw it away, that can be a part of environmental responsibility when you are working with medical items – disposable in this sense means whether or not what the item allows you to do has any real meaning or purpose.

Step 3 : Is the item a part of something that you need to do or want to do in your life?

There is a limit to what you need to be prepared for in life. Chances are you won't need a car that is equipped to handle a sandstorm in the desert.

Buying one encourages production, wastes your resources and creates more generative waste than you can imagine. Always make sure that what you consume or keep in your life as preparation – matches the reality of potential opportunity in your life.

Here are some of the things you can do to reduce waste :

- Print on both sides of the paper to reduce paper wastage.
- Use electronic mail to reach out to people instead of sending paper mail.
- Remove your name from the mailing lists that you no longer want to receive.
- Use cloth napkins instead of paper napkins.
- Avoid using disposable plates, spoons, glasses, cups and napkins. They add to the problem and result in a large amount of waste.
- Avoid buying items that are over-packaged with foil, paper, and plastic. This excess packaging goes to waste.
- Buy durable goods that have an extended warranty. They generally run longer and save landfill space.
- Use refillable pens instead of buying too many.
- Purchase multi-tasking products that perform different kinds of jobs in one.

If something doesn't have a valid purpose or real use anymore, then it adds to your waste. If you're not using any item for a long time, it's time to put it away.

Number 3 deals with the problems created by living in a culture of consumerism. This type of consumption-driven culture also makes fulfilling the second "R" difficult, but it is getting easier to do.

The Second 'R' – Reuse

You may have a box of things you keep that are broken or that you don't have a use for that you hang on to in case you find another use for them; or you may find bargains on old furniture or go trash picking and get things that you can refinish – in either case you are working towards reusing the item. Learning to reuse items, or repurpose them for use different from what they are intended for is essential in the waste hierarchy.

One of the best examples of how this is being done today is the modular construction of homes and office buildings that are being created out of discarded shipping containers.

These large, semi-truck-sized metal containers represent a huge waste problem. Repurposing them as homes and offices saves them from landfills and don't require the additional expenditure of natural resources to melt down and reconfigure the metals used to create them.

You may either reuse those items for your use or donate them so that others can use them.

Reuse the below Items :

Old Jars and Pots: Old jars and pots can be used to store items in the kitchen. They can also be used to store loose items together, such as computer wires.

Tires : Old tires can either be sent to the recycling station or can be used to make tire-swing.

Used Wood : Used wood can be used as firewood or can be used as woodcraft.

Newspaper : Old newspapers can be used to pack items when you're planning to move to another home or store old items.

Envelopes : Old and waste envelopes can be used by children to make short notes.

Waste Paper : Waste paper can be used to make notes and sketches and can be sent to the recycling center when you don't need them anymore.

Old Jeans or T-shirts or any Clothes : These can be turned into bags or tote bags with the help of tutorials available on the internet. At least old clothes can become cleaning rags.

Donate the below Items :

Old Books : Your old books can be used by poor children or can be donated to public libraries.

Old Clothes : Your unwanted clothes can be used by street children or can be donated to charity institutions.

Old Electric Equipment : Old electric equipment can be donated to schools or NGOs so that they can use them.

Rechargeable Batteries : Rechargeable batteries can be used again and again and help to reduce unnecessary wastage as opposed to regular batteries.

Build your compost bin: Use the compost bin to put many waste items like used tea bags, tea leaves and grains, fruit peels, etc. The waste then degrades and turns into compost that helps your plants grow and shine.

Be Creative : Apply DIY on old clothes, bottles, jars, pots, vases, or anything else to convert them into other utilities and use them further.

Repair Damaged Items : Items, when repaired, can be used again without buying replacements.

Second-hand stores: Visit second-hand stores and purchase second-hand goods. Second-hand stores and garage sales can offer usable items in better condition and will save you some money in the process than buying new ones.

The Third 'R' – Recycling

The last stage of the waste hierarchy is recycling. Recycling something means that it will be transformed again into a raw material that can be shaped into a new item. However, there are very few materials on the earth that cannot be recycled.

One of the issues bothering communities that want to become more involved with a recycling effort is that while the relying collection and sorting process may be affordable to implement, there still has to be a facility to receive and transform the discarded waste into a raw material to produce new products.

More progress is being made toward uniting recycling plants with industries that can process the waste material through agreements and incentive credits.

One needs to learn what products can be recycled and what they cannot. By carefully choosing the products that can be recycled, it can be the first step towards efficient recycling.

Buy products from the market that are made up of recycled materials i.e., the product should be environment friendly.

All products are recyclable such as some plastics. There is usually a recycling symbol on the bottom of products to know whether or not it is recyclable.

Buy products that can be recycled like paper, glass, aluminum, plastic, etc. used in the house, school or office. Aluminum can be recycled multiple times.

Invent new ways to recycle different items. However, recyclables need to be gathered systematically and separated from the rest of the trash, ensuring a continuous supply for the recycling process.

Avoid buying hazardous materials that could pose difficulty for you to recycle. Buy non-toxic products whenever possible.

Buy products that have been made from recycled materials.

Using recycled products is as important as recycling items. Recycling can become economically viable only when the recycled products are given significance and put to use.

Use recycled paper for printing or making paper handicrafts.

Benefits of the 'Reduce, Reuse, Recycle' Waste Hierarchy

Since there will be a significant reduction in the amount of waste thrown into the environment, the chances of spreading toxins also decrease. It automatically reduces the levels of greenhouse gas emissions and pollution.

The Waste Hierarchy eliminates the practice of improper waste disposal, i.e., burning waste and trash haphazardly in an uncontrolled manner. Rather it enables us to manage waste in an eco-friendly manner lessening the risk of damage to the environment.

The primary objective of the 3R principle is to lessen the use of newer resources and energy, making more efficient use of resources. It promotes resource efficiency by using the already available resources that are used multiple times, reprocessed, or entirely reduced.

It contributes to more sustainable energy consumption as the resources available on hand are used, and excessive consumption is cut down. It promotes the sustainability of not only energy and resources but also the environment.

It encourages the development of green technology which is a way to create cleaner, safer means of waste disposal while reducing the impact on the environment and all habitats.

It helps increase the use of renewable energy sources like solar, wind, geothermal, etc. as well.

The 3Rs conserve energy and resources and generate jobs in resource management and boost the economy.

Shift from 3Rs to 4Rs

The Fourth 'R' – Recover

Another 'R' is added to the already known 3Rs waste hierarchy and makes it 4Rs. That fourth 'R' is recovering. The 4Rs solutions often come out as a result of industry benchmarking or technological breakthroughs in more innovative companies.

After applying the 3Rs, it may be possible to recover materials or energy out of the waste like electricity, heat, fuel and compost through thermal and biological means, which cannot be reduced, reused or recycled. This can be done by:

Sending treated wastes to a landfill where they will biodegrade away to rejuvenate the land again.

Solid wastes are burned at high temperatures in the incineration method and converted into residue and gaseous products.

Non-recyclable waste is converted into usable heat, electricity, or fuel through varying processes and then becomes a renewable source of energy. Waste to Energy (WtE) helps lessen the need for fossil fuels and reduces emissions of carbon.

Pollution and climate change

Composting, which is a natural biodegradable process, turns organic wastes into nutrient-rich food for plants. It is a slow process, but one of the best ways to turn unsafe organic products into safe compost.

Are there any more 'R's?

Sometimes, two more 'R's can be added to the three basic ones.

Rethink can be added to the start of the list. It means we should think about the way our actions impact the environment.

Recover is sometimes added to the end of the list. It refers to the act of putting waste products to use. For example, decomposing garbage produces methane gas, which can be recovered and burnt to produce energy.

Conclusion :

There are so many ways that your actions, both large and small, can help make your community and the world at large a healthier place. In this series, we'll touch on a few ways you can make changes to "live green."

Are you using the number of resources that you need? Are you disposing of the products you use in a way that doesn't harm the environment? Here are some tips on reducing your consumption, repurposing useful items and recycling others.

Reduce

Use less. Think about how much you consume daily. Can you start by just using less? Watch "The Story of Stuff" with Annie Leonard to learn more about simplifying your life.

Drink from refillable bottles. Disposable plastic bottles use up a lot of resources. Buy a reusable plastic or metal container from which to drink water, and while you're at it, get your daily Starbucks to fix poured directly into an insulated coffee tumbler.

Reuse

Upcycle. Many items you use can be creatively reused. Are you into DIY or crafting? Make some nifty decorative or storage items out of materials such as bottles, boxes or old magazines. Check out San Mateo County's Recycle Works or some Pinterest boards to spark your imagination.

Donate. You might not want that coat anymore, but chances are someone else will. You can donate used clothing, books, kitchen items or furniture in good condition to Goodwill or Salvation Army. They will even pick things up from your home.

Recycle

Put out the bin. In most urban areas, recycling has been made easy for us. If you have a blue bin in your driveway, you probably set out your recycling every week with the trash. If not, check with your local trash collection company or search online to find out what services are available.

Give it at the office. Does your place of employment provide recycling bins for those cans of soda left over from lunch meetings or those papers that got jammed in the printer? If not, see if they can provide recycling containers and disposal for everyone.

Dispose of electronics safely. Many computers, phones, batteries and other devices include toxic materials that can contaminate soil and water if sent to landfills. Take your old gadgets to a retailer such as Best Buy for safe recycling.

Do a bit of Googling. Not sure if something can be recycled? Many things you wouldn't expect can be, from shoes to mattresses to hearing aids. Check out these 25 items you didn't know you could recycle.

*Associate Professor (Zoology)
**Professor
*** Associate Professor (Chemistry)
Raj Rishi Government (Autonomous) College
Alwar, Rajasthan 301001,India.
email : mamta810@gmail.com ;
drhukamsingh63@gmail.com
dr.usingh09@gmail.com

2. A Systematic Review on Prevalence of Dental Fluorosis in Rajasthan

Dr.Neha Goyal[1],
Dr.Neetu Agrawal[2],
Dr.Shiv Singh Dulawat[3]

Abstract
Dental fluorosis is a condition that causes changes in the appearance of tooth enamel. It may result when children regularly consume fluoride during the teeth-forming years, age 8 and younger. Most dental fluorosis in the U.S. is very mild to mild, appearing as white spots on the tooth surface that may be barely noticeable and do not affect dental function. Moderate and severe forms of dental fluorosis, which are far less common, cause more extensive enamel changes. In the rare, severe form, pits may form in the teeth. The severe form hardly ever occurs in communities where the level of fluoride in water is less than 2 milligrams per. Fluorides in drinking water from indigenous rocks and ground water around the mica mines rocks and ground water around the mica mines (Rajasthan has rich sources of mica).

Keywords : Dental fluorosis, ground water, Rajasthan.

Introduction
Regular intake of fluoride containing water can cause fluorosis.Excessive fluoride content in Drinking water contributes to the prevalence of Dental fluorosis. It is a common Public Health problem in endemic areas of India. It is found that it is more prevalent in various districts of Rajasthan. There is three type of fluorosis-1) Dental Fluorosis 2) Skeletal Fluorosis 3) Non Skeletal Fluorosis. In this paper we study about the Dental fluorosis. Concentration of fluoride in drinking water (more than 1.5 mg/ L), food, cosmetics etc. • Low calcium and high alkalinity of drinking water promotes the absorption of F.Dental fluorosis occurs when excessive fluoridated water is ingested during the years of calcification and usage of fluoridated toothpaste before the age of 2 years also contributes to this condition. About 62 million people in India suffer from fluorosis, out of these 80% of the people are

affected by dental fluorosis. Tamil Nadu is one of the 18 states affected by fluorosis in India. Dental fluorosis is characterized by hypo mineralization of tooth enamel. This hypo mineralization is mainly due to insitu toxic effects of fluoride on the ameloblasts of the enamel formation, and not caused by the general effects of fluoride on the calcium metabolism Dental fluorosis is a condition which can easily detected clinically. It is characterized by mottling of enamel, change of colour from yellow to brown to black. In moderate to severe forms tooth may get physically damaged. The clinical findings depends on the degree of dental fluorosis. People with fluorosis are relatively resistant to dental caries Dental fluorosis is endemic in some areas of Rajasthan but it is also found that it is also encountered in nonendemic areas like Alwar. India has 14.1% of the total fluoride deposits on the Earth's crust. It is not surprising, therefore, that fluorosis is endemic in 17 states of India (UNICEF, 1999). In India, the higher concentrations of fluoride in groundwater are associated with igneous and metamorphic rocks and about 62 million people are at risk of fluorosis from drinking high fluoride water. The problem is most pronounced in Andhra Pradesh, Bihar, Gujarat, Madhya Pradesh, Punjab, Rajasthan, Tamil Nadu, and Uttar Pradesh (Husain et al., 2000, 2003, 2004, 2005, 2010, 2012). There are many studies on the distribution of fluoride in groundwater; however, impact assessment studies are still lacking. There is also evidence that the adverse health effects of fluoride are enhanced by a lack of calcium, vitamins and protein in the diet (Zheng et al., 1999). Therefore, fluorosis trends in various social groups were also studied. The Government has introduced some domestic and community-based defluoridation techniques, but they are not accepted by the community. People are still protecting themselves from fluorosis using traditional tactics. A questionnaire was designed to find out the traditional tactics for mitigating fluorosis.

Global and Indian Scenario
International Status
The problem of excessive fluoride in drinking water is prevalent in many parts of the world, and today many millions of people rely on groundwater with concentrations above the World Health Organization guideline value (WHO, 1996). There are >20

developed and developing nations in which fluorosis is endemic (Ayoob & Gupta, 2006). High fluoride concentrations in groundwater are also found in the USA, Africa and Asia (Azbar & Turkman, 2000). The most severe problems associated with high fluoride waters occur in China (Wang et al., 2002); India (Agarwal et al., 2003), Sri Lanka and the Rift Valley countries in Africa. High fluoride groundwaters have been studied in detail in Africa, in particular in Kenya and Tanzania (Moturi et al., 2002). In the early 1980s, it was estimated that 260 million people worldwide (in 30 countries) were drinking water with >1 mg/L of fluoride.

Current Status in India

In India, fluoride was first detected in drinking water at Nellore district of Andhra Pradesh in 1937 (Ayoob & Gupta, 2006). Since then, considerable work has been done in different parts of India to explore the fluoride-laden water sources. At present, it is estimated that fluorosis is prevalent in 17 states of India, indicating that endemic fluorosis is one of the most alarming public health problems of the country, especially in Rajasthan, Madhya Pradesh, Andhra Pradesh, Tamil Nadu, Gujarat, and Uttar Pradesh. At present, endemic fluorosis is thought to affect about one million people (Sneha et al., 2012). Districts known to be endemic for fluoride in various states of India and the ranges of fluoride in drinking water are given in Table 1.

Table 1 Districts showing fluoride concentration >1.5 mg/L in groundwater in India in 2010

State	District	Range
Assam	Goalpara, Kamrup, Karbi Anglong, and Nagaon	1.45–7.8 1.45–7.8
Andhra Pradesh	Adilabad, Anantpur, Chittoor, Guntur, Hyderabad, Karimnagar, Khammam, Krishna, Kurnool, Mahbubnagar, Medak, and Nalgonda	1.8–8.4

Pollution and climate change

State	Districts	Range
Bihar	Aurangabad, Banka, Buxar, Jamui, Kaimur(Bhabua), Munger, Nawada, Rohtas, Supaul	1.7–2.85
Chhattisgarh	Bastar, Bilaspur, Dantewada, Janjgir-Champa, Jashpur, Kanker, Korba, Koriya, Mahasamund, Raipur, Rajnandgaon, and Surguja	1.5–2.7
Delhi	East Delhi, North West Delhi, South Delhi, South West Delhi, West Delhi, Kanjhwala, Najafgarh, and Alipur	1.57–6.10
Gujarat	Ahmadabad, Amreli, Anand, Banaskantha, Bharuch, Bhavnagar, Dohad, Junagadh, Kachchh, Mehsana, Narmada, Panchmahals, Patan, Rajkot, Sabarkantha, Surat, Surendranagar, and Vadodara	1.6–6.8
Haryana	Bhiwani, Faridabad, Gurgaon, Hissar, Jhajjar, Jind, Kaithal, Kurushetra, Mahendragarh, Panipat, Rewari, Rohtak, Sirsa, and Sonepat	1.5–17
Jammu and Kashmir	Doda, Rajauri, and Udhampur	Kashmir 2.0–4.21
Karnataka	Bagalkot, Bangalore, Belgaun, Bellary, Bidar, Bijapur, Chamarajanagar, Chikmagalur, Chitradurga, Davangere, Dharwad, Gadag, Gulburga, Haveri, Kolar, Koppal, Mandya, Mysore, Raichur, Tumkur	1.5–4.4
Kerala	Palakkad, Palghat, Allepy, Vamanapuram, and Alappuzha	2.5–5.7
Maharashtra	Amravati, Chandrapur, Dhule, Gadchiroli, Gondia, Jalna, Nagpur, Nanded	1.51–4.01

Pollution and climate change

Madhya Pradesh	Bhind, Chhatarpur, Chhindwara, Datia, Dewas, Dhar, Guna, Gwalior, Harda, Jabalpur, Jhabua, Khargaon, Mandsaur, Rajgarh, Satna, Seoni, Shajapur, Sheopur, and Sidhi	.5–10.7
Orissa	Angul, Balasore, Bargarh, Bhadrak, Bandh, Cuttack, Deogarh, Dhenkanal, Jajpur, Keonjhar, and Sonapur	0.44-6.0
Punjab	Amritsar, Bhatinda, Faridkot, Fatehgarh Sahib, Firozepur, Gurdaspur, Mansa, Moga, Muktsar, Patiala, and Sangrur	1.52-5
Rajasthan	Ajmer, Alwar, Banaswara, Barmer, Bharatpur, Bhilwara, Bikaner, Bundi, Chittaurgarh, Churu, Dausa, Dhaulpur, Dungarpur, Ganganagar, Hanuman-garh, Jaipur, Jaisalmer, Jalor, Jhunjhunun, Jodhpur, Karauli, Kota, Nagaur, Pali, Rajsamand, Sirohi, Sikar, SawaiMadhopur, Tonk, and Udaipur	1.54–11.3
Tamilnadu	Coimbatore, Dharmapuri, Dindigul, Erode, Karur, Krishnagiri, Namakkal, Perambalur, Puddukotai, Ramanathapuram, Salem, Sivaganga, Theni, Thiruvannamalai, Tiruchirapally, Vellore, and Virudhunagar	1.5–3.8
Uttar Pradesh	Agra, Aligarh, Etah, Firozabad, Jaunpur, Kannauj, Mahamaya Nagar, Mainpuri, Mathura, and Mau	1.5–3.11

Health Effects

Fluorides in drinking water may be beneficial or detrimental depending on their concentration and the total amount ingested. Fluoride is beneficial especially to young children (below eight years of age) for calcification of dental enamel, when present within allowable limits (1.5 mg/L).

Fluoride, being a highly electronegative ion, has an extraordinary tendency to get attracted by positively charged ions like calcium. Hence, the effect of fluoride in mineralized tissues like bone and teeth is of clinical significance as they have the highest amount of calcium and thus attract the maximum amount of fluoride which is deposited as calcium fluorapatite crystals. Tooth enamel is composed principally of crystalline hydroxyapatite. Under normal conditions, when fluoride is present in the water supply, most of the ingested fluoride ions are incorporated into the apatite crystal lattice of calciferous enamel tissue during its formation. The hydroxyl ion is substituted by the fluoride ion because fluorapatite is more stable than hydroxyapatite. The most common health problems associated with excess fluoride in drinking water are dental and skeletal fluorosis. Endemic fluorosis is known to be global in scope, occurring in all continents and affecting many millions of people. Dental fluorosis leads to pitting, perforation and chipping of the teeth, whereas skeletal fluorosis causes severe pains in joints followed by stiffness, which ultimately leads to paralysis. However, recent studies have proved that the health effects of fluoride are not only restricted to dental or skeletal fluorosis but also cause other ailments such as neurological disorders, muscular and allergic manifestations, and gastrointestinal problems, and may also cause lethal diseases like cancer.

Methodology

The fluoride concentration in water was determined electro-chemically, using a fluoride ion selective electrode (APHA, 1991). The two important issues that need to be addressed immediately include the health effects and bottlenecks or problems associated with existing remediation technologies. For collection of data pertaining to evidence, prevalence and severity of dental and skeletal fluorosis, a house-to-house survey was conducted in 63 habitations having fluoride concentration above 5.0 mg/L. For the survey, a questionnaire was designed consisting of information regarding age, sex and dietary habits of individuals. For dental fluorosis, the teeth of individuals and nutritional habits of different age groups and sex were carefully examined in proper daylight. The characteristics of different grades of dental fluorosis are grouped as described by Dean (1942).

- **Normal** The enamel presents a translucent, semi-vitri form type of structure. The surface is smooth, glossy and usually pale creamy white colour.
- **Questionable** Seen in areas of relatively high endemicity; occasional cases are borderline and one would hesitate to classify them as apparently normal or very mild.
- **Very mild** Small, opaque paper-white areas seen scattered irregularly over the labial and buccal tooth surfaces.
- **Mild** The white opaque areas involve at least half of the tooth surface and faint brown stains are sometimes apparent.
- **Moderately** Generally all tooth surfaces are involved and minute pitting is often present on the labial and buccal surfaces. Brown stains are frequently a disfiguring complication.
- **Severe** The severe hypoplasia affect the form of the teeth and stains are widespread, and vary in intensity from deep brown to black.

Using Dean's classification, the Fluoride Index was calculated as: Community fluoride index (CFI) □ □(scores □ no.in each scoregroup) Number of cases examined

For the evidence of skeletal fluorosis, only adult individuals (>21 years) were considered. The grading proposed by Teotia *et al.* (1985) for clinical skeleton fluorosis was considered and used in the present study. The characteristics of different grades are:

- **Grade I** Generalized bone and joint pain.
- **Grade II** Generalized bone and joint pain, stiffness and rigidity of dorso lumber spine and restricted movements at spine and joints.
- **Grade III** Symptoms of grade II with deformities of spine and limbs, knock knees, crippled or bedridden state, kyphosis, invalidism, genu-varum and genu-valgum.

Results and Discussion

The study was carried out in six centrally-located districts of Rajasthan. These districts occupy 76 368 km^2; 1643 habitations in these districts were targeted and of them 72.3% were found to have fluoride above the acceptable limit (1.0 mg/L) of BIS 10500. The distribution of fluoride is shown in Table 3. Fluoride concentration varied from 0.2–23.2 mg/L. The maximum concentration was recorded in samples from Khor habitation of Rani block in Pali district.

Table 3 Fluoride distribution in central Rajasthan.

District	Area (km²)	Habitation Examined	Fluoride Concetration							Concentration Above	
			Min	Max	<1.0	1.0–1.5	1.5–3.0	3.0–5.0	>5.0	Acc. Limit	All. Limits
Ajmer	8 482	190	0.2	15.1	49	37	67	31	6	74.2%	54.7%
Bhilwara	10 455	455	0.2	19.5	132	88	135	53	47	71.0%	51.6%
Jodhpur	22 850	206	0.2	19.7	52	45	72	26	11	74.8%	52.9%
Nagaur	17 683	272	0.4	10.8	52	44	110	38	28	80.9%	64.7%
Pali	12 369	294	0.2	23.2	64	39	119	46	26	78.2%	65.0%
Rajsamand	4 529	226	0.2	6.35	106	59	42	16	3	53.1%	27.0%
Total	76 368	1643	0.2	23.2	455	312	545	210	121	72.3%	53.3%

Acc. Limit – Acceptable limit (1.0 mg/L) and All. Limit – Allowable limit (1.5 mg/L).

In total, 9242 individuals of different age group and sex from 63 habitations were examined for dental fluorosis and 5880 (63.6%) were found to be affected. The maximum number of fluorosis patients was found in Jodhpur district with a maximum CFI of 3.04. The minimum CFI (1.08) recorded was for Ajmer and Bhilwara districts. Table 4 presents the prevalence of various type of dental fluorosis with CFI in the area. In total, 1021 (11%) individuals have severe dental fluorosis of whom 75% have skeletal fluorosis. Skeletal fluorosis was examined for in 4839 individuals of above 20 years age and 1283 (26.5%) were affected. Only 21 (0.4% individual have grade III type skeletal fluorosis. Table 5 presents various grades of skeletal fluorosis in the area.

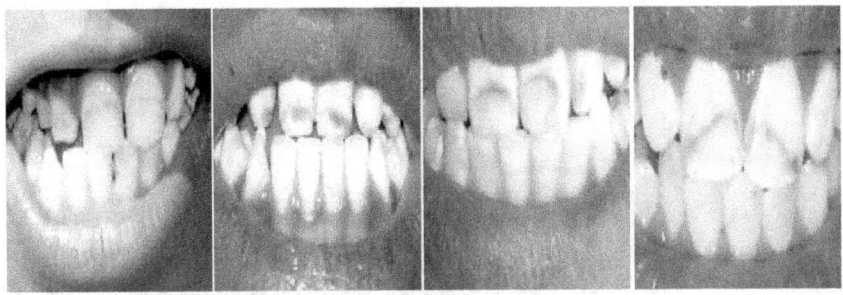

Some dental fluorosis cases are shown in Fig. 1.

Fig. 1 Examples of dental fluorosis in the area.

Table 4 Prevalence of dental and skeletal fluorosis by district.

District	Habitation studied	Individual examined	Dental fluorosis Type I	Type II	Type III	Type IV	Type V	Total	CFI Min	Max
Ajmer	5	590	6.4%	11.7%	18.0%	14.7%	6.9%	57.8%	1.08	1.39
Bhilwara	28	4409	6.4%	12.3%	17.3%	16.3%	10.7%	62.9%	1.08	2.24
Jodhpur	6	835	5.5%	9.6%	16.0%	21.6%	18.7%	71.4%	1.29	3.04
Nagaur	14	2000	5.9%	14.6%	17.3%	15.9%	9.2%	62.8%	1.17	1.91
Pali	8	1203	5.8%	11.6%	16.4%	19.4%	12.3%	65.4%	1.37	1.82
Rajsamand	2	205	12.2%	13.2%	14.6%	11.2%	9.8%	61.0%	1.14	1.65
Total	63	9242	6.2%	12.5%	17.0%	16.9%	11.0%	63.6%	1.08	3.04

Type I: - Questionable, Type II: - Very Mild, Type III: - Mild, Type IV: - Moderate, Type V:- Severe

Table 5 Prevalence of skeletal fluorosis by district.

District	Habitations studied	Skeletal fluorosis Total exam.	Grade I	Grade II	Grade III	Total
Ajmer	5	316	16.8%	8.9%	0.0%	25.6%
Bhilwara	28	2398	17.9%	8.3%	0.6%	26.8%
Jodhpur	6	414	18.4%	10.4%	1.0%	29.7%
Nagaur	14	956	18.4%	8.4%	0.3%	27.1%
Pali	8	651	13.2%	9.2%	0.0%	22.4%
Rajsamand	2	104	19.2%	10.6%	0.0%	29.8%
Total	63	4839	17.4%	8.7%	0.4%	26.5%

Table 6 Prevalence of fluorosis in relation to age and sex

Age (year)	No. of Examination Male	Female	Total	Dental Fluorosis (%) Male	Female	Total	Skeletal Fluorosis (%) Male	Female	Total
5–10	1062	972	2034	803 (75.61%)	680 (69.96%)	1483 (72.91%)	-	-	-
11–20	1238	1131	2369	770 (62.2%)	748 (66.14%)	1518 (64.08%)	-	-	-
21–30	1019	877	1896	706 (69.28%)	492 (56.1%)	1198 (63.19%)	268 (26.3%)	105 (11.97%)	373 (19.67%)
31–40	773	638	1411	458 (59.25%)	374 (58.62%)	832 (58.97%)	211 (27.3%)	183 (28.68%)	394 (27.92%)
41–50	557	411	968	347 (62.3%)	221 (53.77%)	568 (58.68%)	188 (33.75%)	126 (30.66%)	314 (32.44%)
>50	313	251	564	164 (52.4%)	117 (46.61%)	281 (49.82%)	111 (35.46%)	91 (36.25%)	202 (35.82%)
Total	4962	4280	9242	3248 (65.46%)	2632 (61.5%)	5880 (63.62%)	778 (29.23%)	505 (23.2%)	1283 (26.51%)

On categorizing fluorosis in relation to age and sex (Table 6), it was found that dental fluorosis in males is more than that in females. However, the percentage in females falls after the age of 20 years due to the migration of females after their marriage. Males work in the field and consume more water per day and, hence, are more affected by dental and skeletal fluorosis.

Technical Fluoride Mitigation Practices Introduced in the Area

For fluoride mitigation in the area, two technical practices were introduced. The first is based on co-precipitation methods (the Nalgonda technique) and the second is based on adsorption methods using activated alumina. Both techniques are simple and user friendly but no one in the area had adopted either practice.

Co-precipitation Method (Nalgonda Technique)

In this method alum (alumina ferric) is used as a coagulant. The fluoride available in drinking water is adsorbed on the flocs and settles at the bottom of the pot. The supernatant water is collected in another pot and is used for drinking purposes. This method has very low cost and does not require any technical skill, but people in the area have not adopted this technique so far for the following reasons:

- It requires calculation of the alum dose based on alkalinity and fluoride concentration in the raw water. Addition of excessive alum leads to a change in taste due to pH decrease.
- It is a daily job and requires at least 1.5 hours.

Adsorption Method (Activated Alumina Technique)

Granules of activated alumina are used. Activated alumina has a capacity to adsorb fluoride on its surface. After adsorption to a certain level, the further adsorption reduce/stops. When the activated alumina exhausted, it is recharged/regenerated by an alkali and then neutralized by acid. Activated alumina may be regenerated 5–6 times then needs to be replaced. Kits were distributed by the Government at a subsidized rate and NGOs were appointed to establish regeneration centres and IEC. After all these, this technique was still not adopted due to the following reasons:

- Activated alumina is costly; Government subsidised it once but

then it had to be purchased.
- Complications in procedure for monitoring of the activated alumina to assess whether it is still working.
- Regeneration is a difficult task and needs a skilled person. Once the NGO closed the regeneration centre, further use of the technique stopped due to exhaustion of the media.

During the study it was found that these techniques are not effective in the area. However, people were found to use traditional fluoride mitigation practices and by using them they are preventing/delaying fluorosis.

Traditional Fluoride Mitigation Practices used in the Area

To mitigate fluorosis, people in the area have established some rules/practices based on their many years of experience; hence they are well proven. They are based on two sound principles:

- Selection of the least contaminated source for drinking purposes.
- Change in dietary habits.

Selection of Source

In the area with high fluoride in groundwater, people were found to use the least contaminated sources without any knowledge of fluoride availability in water and its ill effects. On further study it was found that there are three ways to select the least contaminated source:

- With the experience of years, local people have categorized sources into ill sources and healthy sources, irrespective of fluoride examination of the water. The ill sources are unsafe and if used result in illness regarding bone deformities. The healthy sources are safe and are used by the community.
- Groundwater sources near to surface water bodies are being used by communities. These sources also provide the least contaminated water.
- In some areas surface water is being used by the community after treatment. Surface water has no fluoride contamination and so the community saved themselves from fluorosis.

Change in Dietary Habits

In the area, people have changed their dietary habits. They were found to use more calcium products in their daily diet. In some area it was established that even eating less but including a bowl of curd in the daily diet was good. Curd has a good quantity of calcium and intake of calcium prevents fluorosis, even if high fluoride water is been used.

Conclusion

In the study a number of habitations were found to have excessive fluoride in groundwater. There is a prevalence of dental fluorosis in a large number of people. However, grade III skeletal fluorosis is rare in the area. This all is due to the well-established traditional practices being used by community in the area.

References
1) Agarwal, M., Rai, K., Shrivastav, R. & Dass, S. (2003) Defluoridation of water using amended clay. *J. Cleaner Produc.* 11, 439–444.
2) APHA (1991) *Standard Methods for the Examination of Water and Wastewater* (17th edn). American Public Health Association. Washington, DC.
3) Ayoob, S. & Gupta, A.K. (2006) Fluoride in drinking water: a. review on the status and stress effects. *Critical Rev. Environ. Sci. Technol.* 36, 433–487.
4) Azbar, N. & Turkman, A. (2000) Defluoridation in drinking waters. 5)*Water Sci. Technol.* 42(1–2), 403–407. BIS (2012) *Drinking Water 6)Standard 10500*. Bureau of Indian Standard, India.
7) Dean, H.T. (1942) The investigation of physiological effects by the epidemiological method. *American Association for the Advancement of Science,* 19, 23–31.
8) Husain, I., Arif, M. & Husain, J. (2012) Fluoride contamination in drinking water in rural habitations of central Rajasthan, India. *Environmental Monitoring and Assessment* 184(8), 5151–5158; doi: 10.1007/s10661-011-2329-7.
9) Husain, J., Husain, I. & Sharma, K.C. (2010) Fluoride and health

hazards: community perception in a fluorotic area of Central Rajasthan (India), an arid environment. *Environmental Monitoring and Assessment* 162, 1–14.
10) Hussain, J., Sharma, K.C. & Hussain, I. (2003) Fluoride distribution in groundwater of Raipur Tehsil in Bhilwara District.
11) *International Journal of Bioscience Reports* 1(3), 580–587.
12) Hussain, J., Sharma, K.C. & Hussain, I. (2004) Fluoride in drinking water and its ill affect on human health: a review. *Journal of Tissue Research* 4(2), 263–273.
13) Hussain, J., Sharma, K.C. & Hussain, I. (2005) Fluoride distribution in groundwater of Banera Tehsil in Bhilwara District, Rajasthan. *Asian Journal of Chemistry*, 17(1), 457–461.
14) Hussain, J., Sharma, K.C., Ojha, K.G. & Hussain, I. (2000) Fluoride distribution in ground waters of Sirohi district in Rajasthan. *Indian Journal of Environment and Eco-planning* 3(3), 661–664.
15) Moturi, W.K.N., Tole, M.P. & Davies, T.C. (2002). The contribution of drinking water towards dental fluorosis: a case study of Njoro Division, Nakuru District, Kenya. *Environ Geochem Health* 24, 123–130.
16) Sneha, J., Yenkie, M.K., Labhsetwar, N. & Rayalu, S. (2012) Fluoride in drinking water and defluoridation of water. *Chem.Rev.* 112, 2454–2466 doi.org/10.1021/cr2002855.
17) Teotia, S.P.S., Teotia, M. & Singh, D.P. (1985) Bone static and dynamic histomorphometry in endemic fluorosis. *Fluoride Research. Studies In Environmental Sciences* 27 347–355.
18) UNICEF (1999) States of the art report on the extent of fluoride in drinking water and the resulting endemicity in India. *Report by fluorosis and rural development foundation for UNICEF.* New Delhi: UNICEF.
19) US Public Health Service (1962) *Drinking Water Standards.* Department of Health Education and Welfare, Washington, DC. Wang, W., Li, R., Tan, J., Luo, K., Yang, L., Li, H. & Li, Y. (2002) Adsorption and leaching of fluoride in China. *Fluoride* 35, 122–129.
20) WHO (World Health Organization) (1996) *Guideline for*

Drinking Water Quality. Geneva: World Health Organization.
21) Zheng, B.S., Dingm, Z.H., Huang, R.G., Zhu, J.M., Yu, X.Y. & Wang, A.M. (1999) Issues of health and disease relating to coal use in southwestern China. *International Journal of Coal Geology* 40(2–3), 119–132.

[1]Dept of chemistry,
Agrasen Mahila Mahavidhyalay, Kherli (Alwar)
Rajasthan, India
[2] Dept of chemistry,
Agrawal PG College, Jaipur,
Rajasthan, India
[3] Dept of Chemistry,
B.N. University, Udaipur,
Rajasthan, India
email : agrawalneetu05@gmail.com

3. Climate Change and its Effect on Environment

Nirmala Bansal

Introduction

Over the past century global average temperature has increased byo8 ■. Scientific alarm about the buildup of the so called „Green House Gases" & the theory that these could lead to global climate change was sounded. Earth climate is influenced usually by the primary 18 kilometers.

This layer contains 78% Nitrogen, 21% Oxygen & 1% other gases like Carbondioxide (CO_2), Methane (CH_4), Ozone (O_3) CfC_s, nitrous oxide (N_2O) etc. These 1% gases in the atmosphere are called green house gases.

Fossil, fuel burning, deforestation & the release of Industrial chemicals are rapidly heating the earth to temperatures. Most important in CO_2. Such gases absorbs infrared radiations emitted by the earth surface and act as blankets over the surface keeping it warmer them it would be the concentrations of CO_2 and methane have increased by 36% & 148% in last two decades. Therefore the global warming is the gradual increase of Earth"s surface temperature due to Green house gases like dioxide emissions.

Impact :

Limiting global heating & climate change is the central environmental challenge of our time.

Weather Change : Rising temperature & have several effects on the weather i.e.

1. Sea level will rise all over the earth. Major weather patterns i.e. the tropical mansoons & Jet streams are altered.
2. It increase evapotranspiration in which water is evaporated from Soil, Plants, water bodies. This effects frequency & intensity of droughts. A hotter atmosphere can hold more water uapours than previous years i.e. 4% approx. this increases the chance of maximum rainfall events.

Pollution and climate change

Shrinkage of Polar Ice Caps : The whole world may face threats of fast shrinkage of polar ice melting, fast rise within the water level, danger for species like polar bears etc. USA, UK, lower Islands can be plagued by cold waves, heavy snow falls & stroms due to the shifting & melting of largest ice sheets in the Atlantic Sea.

Agriculture : Heating can effect agriculture. When temperature increases by 1.5°C– 2.5°C agriculture production would decline by 4.04 – 6.73%. If annual Precipitation decreases by 10 – 20% agricultural Production would decrease by 0.965% – 1.93% approx. Plants are the most important source of food and as a result food storage black marketing may occur; which may cause war & conflicts in some countries.

Effect on Animals, Bird Life & Plants : Climate change & warming is impacting every living being on planet i.e. birds are laying their eggs earlier each year, flowers are blooming earlier & animals who are hibernating wake up earlier; Rising sea level ($\sim 50 cm$), sea turtles will lose their nesting grounds and result in extinction. This is especially true for migratory wildlife & plants including mangroves, sea grass and coral and also cause the reduction in the population of fish. Totally distribution of animals is affected, extinction of plants increases, birds are migrating & arriving at their nesting grounds earlier.

Spread of Diseases : Increase in temperature can effect the health of human & therefore the diseases they exposed to Bird flu, Cholera, Ebola, Plague & tuberculosis are just few of the diseases likely to spread & get worse as a result of climate change according to (WCS) wild life conservation Society with the rise in the rainfall, water born diseases like malaria, west Nile, Cholera, Lyme disease and dengue are likely to spread. High air temperature increases the Ozone concentration at ground level. It damages lung tissues & aggravates asthma & other breathing problems.

Solutions to Global Warming : There are measures if implemented soon can Reduce the Social, economic, environmental & Political impact of changing climate. In fact there is no single solution to heating, which is primarily an issue of an excessive amount of heattrapping greenhouse emission (CO_2), methane and inhalation anesthetic within the atmosphere. The technologies and approaches outlined below are all needed to bring down the emissions of those gases by a minimum of 80 percent by mid-century. So as to determine how they're best deployed in each region of the world:

Boosting Energy Efficiency : The energy accustomed power, heat, and funky our homes, businesses, and industries is that the single largest contributor to warming.

A method to Minimize greenhouse emission and help to manage the atmospheric phenomenon is to limit our energy consumption. Energy efficiency technologies allow us to use less energy to induce the identical or higher level of production, service, and luxury.

Green Transportation : Transportation could be a major contributor to gas emissions. So it's the prime target for reducing pollution and obtaining sustainable environment. This ends up in Green Transportation, which implies any quite transportation practice or vehicle that's eco-friendly and doesn't have any negative impact on the immediate environment as so far.

Renewable Energy Sources : Renewable energy sources like solar, wind, geothermal and bioenergy are available round the world. Renewable energy create clean energy and are in use round the world for several years with low carbon emissions. Renewable technologies are deployed quickly, are increasingly cost-effective, and make jobs while reducing pollution. Govt. is also making policies on it.

Phasing out Fossil Fuels : Burning of fossil fuels like wood or coal produce more carbon emissions than other product. Phasing out coal burning power plants and not burning fossil fuels directly will reduce dependence of fossil fuels.

Managing Forests and Plant Trees : As CO_2 as is that the most vital greenhouse emission, planting trees and other plants can slow or stop heating. one tree can absorb up to 48 pounds of CO_2 a year; one acre of them consumes 2.6 tons. They use carbon to make their own tissues and return a number of it to the soil during a process called sequestration. Deforestation may be a large contributor to heating planting new trees, can be the solution.

Exploring Nuclear Energy : Because nuclear energy leads to few heating emissions, an increased share of atomic energy can help in reducing heating. But nuclear technology poses serious threats to our security.

Developing and Deploying new low-Carbon and Zero-Carbon Technologies : Research into and developments of the following generation of low-carbon technologies are going to be critical to deep mid-century reductions in warming. Current research on battery technology, new materials for solar cells and other innovative areas can help.

Water Conservation : Significant amounts of energy are used while purifying and distributing water, which contributes to gas emissions. Saving water reduces the number of energy used. At home, put off water immediately whenever you are not using it. In your yard, landscape with plants and grasses that need less water, and capture rainwater for irrigating.

Reducing Waste and Recycling : the assembly of garbage contributes to heating both directly and indirectly. Decomposing waste in landfills produces CO_2, methane and other greenhouse gases. When the waste is burnt, it releases toxic gases which lead to warming. Reducing your consumption patterns and reusing items whenever possible minimizes your carbon footprint, since fewer new items must be made. Recycling metal, plastic, glass, paper and other recyclable items can help to cut back heating.

Conclusion : Global warming is Phenomenon of climate change with contributes to changing patterns of extreme weather across the globe from longer & hotter heat waves to heaver rains. No nation whether large or small, wealthy or poor, developed or developing modern or ancient can escape the impact of Global warming. We continue our daily routine of emitting tons and a lot of green house gases helping the planet heat up or we shall take the measures now to avoid a catastrophe. Our govt. is also taking measures regarding it. We cannot be late now. We have to reduce Carbon emission, work on conservations & right use of resources. Over all that our earth is "Sick", we humans nead to "heal" the earth. Future global warming will depend on future human actions. We must figure from today & hitch our hands with co-operation. We shall not subject ourselves to doom and destruction.

Humans were the one who changed the world & now it is time for the human to change themselves. It"s the time now that we join our hands together & combat this global threat effectively keeping visible our own security & our future generations existence.

References :
Jain Sharad K & Kumar Vijiy (2012). Trend analysis of rainfall & temperature data for
India current Science Vol. 102 No. 1 Jan 2012.
Barat Raj Singh. 2015, "Global Warming: Causes. Impacts and Remedies." Chapter -3
ISBN-13: 9789535120438, In Tech 2015.
Gillis, Justin (28 November 2015). "Short Answers to Hard Questions About Climate
Change". The New York Times.
EPA (2007). "Recent Climate Change: Atmosphere Changes". Climate Change Science
Program. United States Environmental Protection Agency.

Assistant Professor (Chemistry)
SRP Govt. College Bandikui (Daussa),
Rajasthan

4. Novel 1,3,4-oxadiazole Derivatives : Design, Characterization and Biological evaluation

Jyoti Sharma[1], *Nidhi Agarwal[2]

Abstract

Oxadiazoles are the heterocyclic compounds having Nitrogen and oxygen as heteroatoms with carbon. Oxadiazoles are of 4 types on the basis of position of heteroatoms in five membered ring. Among which 1,3,4-oxadiazole has been studied much in last decades due to its excessive antimicrobial activities. In the present work we have synthesized some N-substituted derivatives of 1,3,4-oxadiazole with the combination of substituted acid hydrazide and aryl isothiocynates in the presence of KI/I_2 in alkaline solution of NaOH. The structure of synthesized compounds was demonstrated by ^1HNMR, IR , $^{13}CNMR$, Mass spectroscopy and elemental analysis. These compounds of series 5(a-e) were evaluated for antibacterial and antifungal activity against bacteria E. coli & S. aureus and fungus C. albicans & A. niger. Newly formed compounds revealed significant antimicrobial activity. The present research can support the new researchers for further characterization and evaluation of different derivatives of 1,3,4-oxadiazole compounds for the development of new useful antimicrobial agents.

Keywords : Heterocyclic Chemistry, 1,3,4-oxadiazole, Antibacterial activity, Antifungal Activity, Isothicyantes,

1. Introduction

According to the investigation findings, heterocycles make up 69 percent of drugs in the late stages of development or are already on the market. Heterocyclic compounds are essential to life in various ways and may be widely distributed across nature.[1-2] Oxadiazoles are a subclass of heterocyclic compounds containing the hetero-atoms of nitrogen and oxygen and carbon heteroatoms. There are four distinct types of oxadiazoles, and each one is distinguished from the others by the position of the heteroatoms inside the five-membered ring. Therefore, medicinal chemistry experts have investigated 1,3,4-oxadiazole's antitubercular[3-5] capabilities as well as its anti-inflammatory[6-7], anticonvulsant, antifungal, and antibacterial activities[8] have been of interest. A recent thorough study has demonstrated that 1,3,4-oxadiazole can play a pivotal role as a precursor chemical in producing innovative and improved compounds that have favourable biological characteristics and might be used in various applications. Nesapidil (an anti-arrhythmic)[9], Fenadiazole (a hypnotic)[10], Raltegravir (an HIV replication inhibitor)[11], Tiodazosin (an antihypertensive)[12], and Zibotentan[13] are some of the medications that are currently available and contain 1,3,4-oxadiazole. Zibotentan is also one of the medications that treat hypertension (anticancer).

Bacteria are a microscopically small group that belongs to the kingdom of plants. They are devoid of chlorophyll and have no colour.[14] Bacteriology is the branch of the life sciences that focuses on the investigation of microorganisms, specifically bacteria. The Gram stain, which was created by the Danish physician Christiaan Gram, is the basis for the categorization of all bacteria into one of two categories, which are referred to alternately as "Gram-positive" and "Gram-negative" (1884)[15-17]. Different from other types of single-celled or very simple multicellular organisms are fungi. Although they may be discovered virtually everywhere, the vast majority choose to make their home on land, more specifically in the soil or plant matter, as opposed to the open water or a body of freshwater. Various categories can be used to classify fungi, including how they reproduce, whether or not they have fruiting bodies, how they are structured, and the spores they produce[18]. Bacteria capable of causing disease in people or animals are called pathogens. Compounds are said to possess antimicrobial properties if they can prevent the growth of various microorganisms, including bacteria, fungus, viruses, amoebas, and so on, or eliminate them. Chemicals that kill microorganisms can be categorized as antibacterial, antifungal, antiviral, antiprotozoal, or antitubercular, depending on how well they work against bacteria, fungus, viruses, amoebas, and other types of microorganisms[19]. Antitubercular drugs are particularly effective against tuberculosis. This category includes such well-known drugs as penicillin, cephaloridine, colistin, gentamycin, streptomycin, methicillin, clotrimazole, miconazole, fluconazole, and ketoconazole, as well as some additional pharmaceuticals[20]. Bactericidal drugs are preferable to bacteriostatic medications with a delayed beginning of the action, such as chloramphenicol, doxycycline, and sulphonamide. Bactericidal agents kill bacteria outright. [21]

Infectious bacteria are known to develop resistance to antimicrobial medicines at a rate that much exceeds the rate at which new therapies are developed. Therefore, it is possible that in the future, research into antimicrobial therapy will centre on either the development of techniques to battle resistance or the discovery of novel ways to treat illnesses.

In the current study, we created several N-substituted derivatives of 1,3,4-oxadiazole by using aryl isothiocyanates[22] and substituted acid hydrazides in the presence of KI/I_2 in an alkaline solution of NaOH. These derivatives were then tested for their capacity to prevent the development of fungus and bacteria.

2. Experimental Section:

Each of the chemicals and the reagents was procured from their retail outlets. The open capillary technique was utilized to determine the melting points, and the data obtained were not modified in any way. Iodine vapours were used to analyze the spots, while the reaction was observed using silica gel-coated glass plates for thin-layer chromatography. We generated ^1HNMR and $^{13}CNMR$ spectra by employing a Bruker Advance Neo 500MHz NMR spectrophotometer, using DMSO as the solvent and tetramethyl silane as the internal standard. To capture the IR spectra, an F. T. InfraRed Spectrophotometer Model RZX was utilized (Perkin Elmer). An LC-MS Spectrometer Model Q-ToF Micro Waters was utilized to calculate the mass spectra. The elements analysis has to be carried out for the FLASH 2000 CHN Elemental Analyzer to work properly. Therefore, it was decided that Thermo Scientific would be a good fit for the job. Numerous instances include the use of the letters s (singlet), d (doublet), dd (double doublet), t (triplet), and m (multiple) to denote the different types of multiplication. The 1,3,4-oxadiazole compounds with N substitutions were created following the Fig 1.

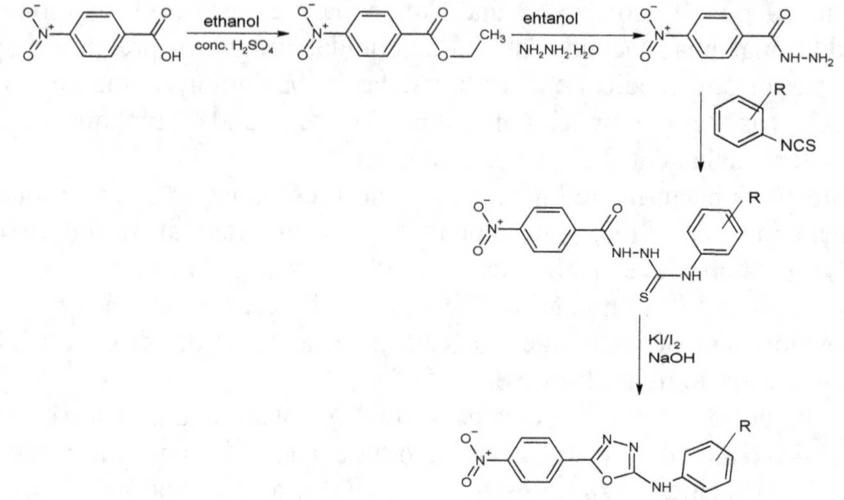

R= (a) H, (b) 2,6- CH$_3$, (c) 4-CH$_3$, (d) 4-Cl, (e) 4-OCH$_3$

Fig 1.: Synthetic scheme of the N-substituted-1,3,4-oxadiazole derivatives

2.1: General steps for synthesizing N-substituted-1,3,4-oxadiazole derivatives :

2.1.1: Step-1 : Ethyl-4-nitrobenzoate synthesis :

Concisely and drop by drop 4-nitrobenzoic acid (0.0623 mol) **(1)** that had been dissolved in 12 ml ethanol was used to form the reaction mixture. Additionally, 0.5 ml of sulphuric acid was added to the mixture. After heating to an internal temperature of between 40 and 50 degrees Celsius, the mixture was allowed to reflux for 7 hours. After an excessive amount of alcohol was distilled and cooled by submerged in ice water. We made use of TLC so that we could monitor how far along the reaction was progressing as it was taking place. After the cooling process, the combination was finished; it was filtered, washed with water, and recrystallized with ethanol. The rest of the preparations began as soon as the operation was over.

Yield: 90%; b.p. 186.3°C ; ^{1}H NMR: δ 1.24 (3H, t, J = 7.1 Hz CH$_3$), 4.25 (2H, q, J = 7.1 Hz CH$_2$), 6.93 (2H, ddd, J = 8.6, 1.2, 0.5 Hz Ar-H), 8.07 (2H, ddd, J = 8.6, 1.8, 0.5 Hz Ar-H); IR(KBr, cm^{-1}): 3178.8(Ar-H stretch.), 1693.3(C=O stretch.), 1302.3(C-NO$_2$ stretch), 1054.5(C-O stretch); MS m/z: 195.05(M$^+$). Anal. Calcd. For C$_9$H$_9$NO$_4$ C-55.39; H-4.65; N-7.18; O-32.79; Found: C-55.90; H-

4.11; N-8.90; O-32.67%.

2.1.2: Step-2: Synthesis of 4-nitro benzohydrazide :
Compounds **3** was produced as a result of a reaction in which ethyl 4-nitro benzoate (**2**) (0.0723 mol) and hydrazine hydrate (6 ml) were refluxed together for nine hours at temperatures ranging from 30 to 40 degrees Celsius. The liquid was allowed to reach room temperature before being poured onto the crushed ice below while continuously agitated. After allowing the temperature to return to normal, this was carried out. TLC was utilized to keep track of the reaction being observed and maintain track of the reaction being monitored. After filtering and washing the residue with water, ethanol was employed to recrystallize it so that the final product would have crystals. The result of this technique was a material that had a crystalline structure.
Yield: 89%; m.p. 150-152°C ; ^{1}H NMR: δ 6.94 (2H, ddd, J = 8.6, 1.2, 0.5 Hz Ar-H), 8.11 (2H, ddd, J = 8.6, 1.8, 0.5 Hz Ar-H); IR(KBr, cm^{-1}): 3442.1(N-H stretch.), 3183.4(Ar-H stretch.), 1707.6(Amide C=O stretch), 1387.7(C-NO$_2$ stretch); MS m/z: 181.04(M$^+$). Anal. Calcd. For $C_7H_7N_3O_3$ C-46.41; H-3.89; N-23.20; O-26.50; Found: C-45.90; H-4.01; N-23.21; O-25.09%.

2.1.3: Step-3: Synthesis of 2-(4-nitrobenzol)-*N*-substituted phenyl hydrazine-1-carbothioamide :
After adding 24 millilitres of ethanol to the combination that included Compound 3 and the substituted phenyl isothiocyanates (a-e), the mixture was left to reflux over a water bath for a total of two hours. The product could precipitate once the solvent was concentrated, and after being dried, filtered, and crystallized from methanol for a second time, it was finally ready for use. Compounds **4(a-e)** were produced as a result of this method, and their identities were determined as follows:

2-(4-nitrobenzoyl)-*N*-phenylhydrazine-1-carbothioamide (4a):
Yield: 92%; m.p. 110-112°C ; ^{1}H NMR: δ 6.95 (1H, tt, J = 7.8, 1.2 Hz N-H), 7.18-7.40 (4H, 7.25 (dddd, J = 8.2, 7.8, 1.4, 0.5 Hz Ar-H), 7.51-7.65 (3H, Ar-H), 8.00 (2H, dddd, J = 8.5, 1.9, 1.5, 0.4 Hz N-H). IR(KBr, cm^{-1}): 3406.5(N-H stretch.), 3194.5(Ar-H stretch.), 1693.8(Amide C=O stretch), 1516.6(C=S stretch), 1294.8(C-NO$_2$ stretch); MS m/z: 316.06(M$^+$). Anal. Calcd. For $C_{14}H_{12}N_4O_3S$: C-

53.16, H-3.82, N-17.71, O-15.17, S-10.14; Found: C-54.90, H-3.22, N-16.09, O-15.09, S-9.99%

N-(2,6-dimethylphenyl)-2-(4-nitrobenzoyl)hydrazine-1-carbothioamide (4b) :

Yield: 85%; m.p. 120-122°C ; ^{1}H NMR: δ 2.15 (6H, s CH$_3$), 6.83 (1H, t, J = 7.8 Hz N-H), 6.98 (2H, dd, J = 7.8, 2.5 Hz Ar-H), 7.51-7.65 (3H, 7.57 (tt, J = 7.5, 1.5 Hz Ar-H).; IR(KBr, cm^{-1}): 3391.0(N-H stretch.), 3185.3(Ar-H stretch.), 3037.2(C-H aliph.), 1687.6(Amide C=O stretch), 1518.8(C=S stretch), 1385.7(C-NO$_2$ stretch); MS m/z: 344.09(M$^+$). Anal. Calcd. For C$_{16}$H$_{16}$N$_4$O$_3$S: C-55.80, H-4.68, N-16.27, O-13.94, S-9.31; Found: C-53.08, H-4.44, N-14.23, O-15.09, S-9.22%.

N-(4-methylphenyl)-2-(4-nitrobenzoyl)hydrazine-1-carbothioamide (4c) :

Yield: 65%; m.p. 162-164°C ; ^{1}H NMR: δ 2.21 (3H, s CH$_3$), 7.01-7.15 (4H, 7.07 (ddd, J = 8.1, 1.3, 0.6 Hz Ar-H), 7.09 (ddd, J = 8.1, 1.1, 0.6 Hz)), 7.51-7.65 (3H, 7.57 (tt, J = 7.5, 1.5 Hz N-H),; IR(KBr, cm^{-1}): 3390.9(N-H stretch.), 3184.9(Ar-H stretch.),3038.8(C-H aliph.), 1690.6(Amide C=O stretch), 1218.5(C=S stretch), 1299.3(C-NO$_2$ stretch); MS m/z: 330.07(M$^+$). Anal. Calcd. For C$_{15}$H$_{14}$N$_4$O$_3$S: C-54.53, H-4.27, N-16.96, O-14.53, S-9.71; Found: C-45.90, H-4.01, N-15.21, O-13.09, S-9.21%.

N-(4-chlorophenyl)-2-(4-nitrobenzoyl)hydrazine-1-carbothioamide (4d) :

Yield: 60%; m.p. 210-212°C ; ^{1}H NMR: δ 7.35-7.65 (7H, 7.41 (ddd, J = 8.1, 1.6, 0.5 Hz Ar-H), 7.49 (ddd, J = 8.1, 1.3, 0.5 Hz), 7.57 (tt, J = 7.5, 1.5 Hz).; IR(KBr, cm^{-1}): 3410.0(N-H stretch.), 3185.7(Ar-H stretch.), 1699.0(Amide C=O stretch), 1220.0(C=S stretch), 1285.8(C-NO$_2$ stretch) 1048.2(C-Cl stretch); MS m/z: 350.02(M$^+$). Anal. Calcd. For -C$_{14}$H$_{11}$ClN$_4$O$_3$S: C-47.94, H-3.16, N-15.97, O-13.68, S-9.14, Cl-10.11; Found: C-45.22, H-3.31, N-15.41, O-13.09, S-9.33, Cl-9.90%.

N-(4-methoxyphenyl)-2-(4-nitrobenzoyl)hydrazine-1-carbothioamide (4e) :

Yield: 75%; m.p. 213-215°C ; ^{1}H NMR: δ 3.76 (3H, s CH$_3$), 6.69 (2H, ddd, J = 8.8, 2.7, 0.5 Hz Ar-H), 7.20 (2H, ddd, J = 8.8, 2.2, 0.5 Hz, Ar-H), 7.51-7.65 (3H, 7.57 (tt, J = 7.5, 1.5 Hz); IR(KBr, cm^{-1}):

3407.7(N-H stretch.), 3007.9(C-H stretch), 3190.9(Ar-H stretch.), 1699.4(Amide C=O stretch), 1222.3(C=S stretch), 1291.3(C-NO$_2$ stretch); MS m/z: 346.07(M$^+$). Anal. Calcd. For: $C_{15}H_{14}N_4O_4S$: C-52.02, H-4.07, N-16.18, O-18.48, S-9.26; Found: C-52.00, H-4.09, N-15.21, O-18.33, S-9.08%.

2.1.4: Step 4: Synthesis of N-substituted-1,3,4-oxadiazole derivatives(5a-e) :

To get a transparent solution, Sodium hydroxide in aqueous form was added to the mixture that already included hydrazine carbothioamides 4a-e and ethanol (50 millilitres). Next, I$_2$ was gradually added to the solution containing 10 % KI while continually agitated. This was done to ensure that the colour of the I2 remained stable at room temperature. This I$_2$/KI solution was maintained at 10 degrees Celsius. The mixture of hydrazine carboxamide, NaOH, and ethanol was then given this combination to add to it, and it was refluxed for five hours at temperatures ranging from 30 to 40 degrees Celsius. Finally, the refluxing mixture was recrystallized with ethanol after it was cooled, filtered, and thoroughly washed with water.

Compounds **5a -5e** were produced as a result of following the technique above, and they were subsequently identified as follows:

5-(4-nitrophenyl)-*N*-phenyl-1,3,4-oxadiazol-2-amine (5a)

Yield: 95%; m.p. 223-225°C ; ^{1}H NMR: δ 7.04 (1H, tt, J = 7.9, 1.2 Hz, N-H), 7.28 (2H, dddd, J = 8.2, 7.9, 1.7, 0.5 Hz, Ar-H), 7.48 (2H, dddd, J = 8.2, 1.3, 1.2, 0.5 Hz, Ar-H), 7.62-7.83 (4H, Ar-H); ^{13}C NMR: δ 117.7 (2C, s), 119.9 (2C, s), 122.8 (1C, s), 127.8 (1C, s), 128.2 (2C, s), 129.4 (2C, s), 137.2 (1C, s), 139.5 (1C, s), 158.9 (1C, s), 163.2 (1C, s); IR(KBr, cm^{-1}): 3404.1(N-H stretch.), 3209.8(Ar-H stretch.), 1551.5(C=C stretch), 1455.3(C=N stretch), 1339.1(C-O-C stretch), 1278.7(C-NO$_2$ stretch), 1239.3(N-N stretch); MS m/z: 282.07(M$^+$). Anal. Calcd. For: $C_{14}H_{10}N_4O_3$: C-59.57, H-3.57, N-19.85, O-17.01; Found: C-60.01, H-3.19, N-18.31, O-16.33%.

***N*-(2,6-dimethylphenyl)-5-(4-nitrophenyl)-1,3,4-oxadiazol-2-amine (5b)**

Yield: 80%; m.p. 214-216°C ; ^{1}H NMR: δ 2.21 (6H, s, CH$_3$), 6.82 (1H, t, J = 7.9 Hz, N-H), 7.00 (2H, dd, J = 7.9, 2.7 Hz), 7.62-7.83 (4H, 7.68 (ddd, J = 8.3, 1.5, 0.4 Hz), 7.77 (ddd, J = 8.3, 1.7, 0.4 Hz)

Ar-H); ^{13}C NMR: δ 17.7 (2C, s), 117.7 (2C, s), 122.8 (1C, s), 128.0 (1C, s), 129.0 (2C, s), 129.2 (2C, s), 129.4 (2C, s), 137.2 (1C, s), 139.5 (1C, s), 158.9 (1C, s), 163.2 (1C, s).; IR(KBr, cm^{-1}): 3390.1(N-H stretch.), 3184.5(Ar-H stretch.), 3044.1(C-H stretch), 1556.6 (C=C stretch), 1478.5(C=N stretch), 1317.0(C-O-C stretch), 1278.1(C-NO$_2$ stretch), 1225.5(N-N stretch); MS m/z: 310.10(M$^+$). Anal. Calcd. For: $C_{16}H_{14}N_4O_3$: C-61.93, H-4.55, N-18.06, O-15.47; Found: C-60.99, H-4.21, N-17.90, O-15.33%.

N-(4-methylphenyl)-5-(4-nitrophenyl)-1,3,4-oxadiazol-2-amine (5c)

Yield: 65%; m.p. 233-235°C ; ^{1}H NMR: δ 2.18 (3H, s, CH$_3$), 7.02-7.23 (4H, 7.09 (ddd, *J* = 8.1, 1.4, 0.5 Hz), 7.17 (ddd, *J* = 8.1, 1.3, 0.5 Hz) Ar-H), 7.62-7.83 (4H, 7.68 (ddd, *J* = 8.3, 1.5, 0.4 Hz), 7.77 (ddd, *J* = 8.3, 1.7, 0.4 Hz) Ar-H); ^{13}C NMR: δ 21.3 (1C, s), 117.7 (2C, s), 117.9 (2C, s), 122.8 (1C, s), 129.4 (2C, s), 129.6 (2C, s), 137.2 (1C, s), 139.5 (1C, s), 141.5 (1C, s), 158.9 (1C, s), 163.2 (1C, s); IR(KBr, cm^{-1}): 3403.1(N-H stretch.), 3187.6(Ar-H stretch.), 3040.9(C-H stretch), 1563.5(C=C stretch), 1451.4(C=N stretch), 1340.5(C-O-C stretch), 1277.3(C-NO$_2$ stretch), 1234.4(N-N stretch); MS m/z: 296.09(M$^+$). Anal. Calcd. For: $C_{15}H_{12}N_4O_3$: C-60.81, H-4.08, N-18.91, O-16.20; Found: C-59.44, H-4.09, N-17.45, O-15.90%.

N-(4-chlorophenyl)-5-(4-nitrophenyl)-1,3,4-oxadiazol-2-amine (5d)

Yield: 80%; m.p. 200-202°C ; ^{1}H NMR: δ 7.42 (2H, ddd, *J* = 8.1, 1.6, 0.5 Hz, Ar-H), 7.62-7.83 (6H, 7.69 (ddd, *J* = 8.3, 1.5, 0.4 Hz), 7.69 (ddd, *J* = 8.1, 1.6, 0.5 Hz), 7.77 (ddd, *J* = 8.3, 1.7, 0.4 Hz) Ar-H); ^{13}C NMR: δ 117.7 (2C, s), 120.5 (2C, s), 122.8 (1C, s), 128.9 (2C, s), 129.4 (2C, s), 133.7 (1C, s), 137.2 (1C, s), 139.5 (1C, s), 158.9 (1C, s), 163.2 (1C, s); IR(KBr, cm^{-1}): 3396.2(N-H stretch.), 3192.7(Ar-H stretch.), 1559.2(C=C stretch), 1472.7(C=N stretch), 1316.4(C-O-C stretch), 1283.2(C-NO$_2$ stretch), 1239.3(N-N stretch), 1048.6(C-Cl stretch); MS m/z: 316.03(M$^+$). Anal. Calcd. For: $C_{14}H_9ClN_4O_3$: C-53.09, H-2.86, N-17.69, O-15.16, Cl-11.19; Found: C-52.44, H-2.09, N-16.27, O-15.63, Cl-10.99%.

N-(4-methoxyphenyl)-5-(4-nitrophenyl)-1,3,4-oxadiazol-2-amine (5e)

Yield: 73%; m.p. 234-236ºC ; ^{1}H NMR: δ 3.74 (3H, s, CH$_3$), 6.64 (2H, ddd, J = 8.8, 2.7, 0.5 Hz, Ar-H), 6.88 (2H, ddd, J = 8.8, 1.7, 0.5 Hz), 7.62-7.83 (4H, 7.69 (ddd, J = 8.3, 1.5, 0.4 Hz), 7.77 (ddd, J = 8.3, 1.7, 0.4 Hz) Ar-H); ^{13}C NMR: δ 56.0 (1C, s), 114.5 (2C, s), 117.7 (2C, s), 120.5 (2C, s), 122.8 (1C, s), 129.4 (2C, s), 137.2 (1C, s), 139.5 (1C, s), 158.9 (1C, s), 159.8 (1C, s), 163.2 (1C, s); IR(KBr, cm^{-1}): 3403.1(N-H stretch.), 3205.6(Ar-H stretch.), 3007.1(C-H stretch), 1559.3(C=C stretch), 1474.9(C=N stretch), 1320.1(C-O-C stretch), 1178.2(C-NO$_2$ stretch), 1226.7(N-N stretch); MS m/z: 312.08(M$^+$). Anal. Calcd. For: C$_{15}$H$_{12}$N$_4$O$_4$: C-57.69, H-3.87, N-17.94, O-20.49; Found: C-57.55, H-3.88, N-16.89, O-19.87%.

3. Biological Evaluation :

The disc-diffusion technique[23] was employed to assess the newly synthesized compounds 5(a-e) for their potential to prevent bacterial growth in vitro. In addition, these tests were carried out to determine the capabilities of the substances. Depending on the specific type of pathogen used, 50 or 100 µg/ml was the concentration used for the pathogens. This collection of pathogens comprised both Gram-positive and Gram-negative strains of the bacterium E. coli and S. aureus. Using the technique mentioned above, the size of the zone of inhibition in the nutritious agar media was calculated in millimetres. To produce the concentration solutions for the newly synthesized compounds, the compounds were first allowed to dissolve in DMF before being mixed. When the common antibiotic ofloxacin was used, it revealed an inhibitory zone of 26 millimetres for E. coli and 25 millimetres for S. aureus at a concentration of 100 milligrams per millilitre. The findings are outlined in Table 1, which may be seen further down on this page.

In a manner analogous to this, the disc diffusion technique was utilized to evaluate the newly synthesized compounds 5(a-e) for their ability to inhibit fungal growth. Aspergillus niger and Candida albicans, separate fungus species, were tested at levels of fifty and one hundred g/ml, respectively. In the experiment, which used the medication fluconazole as a point of comparison, the zone of inhibition for A. niger measured 23 millimetres, while the zone for C. Albicans measured 25 millimetres. The findings are summarized in Table 1, which may be seen below.

Table 1. Antibacterial and Antifungal activity of synthesized compounds

Compound	Zone of Inhibition							
	S. aureus (Gram+ve)		E.coli(Gram -ve)		A. niger		C. albicans	
Conc. µg/ml	50 µg	100 µg	50 µg	100 µg	50 µg	100 µg	50 µg	100 µg
5a	17	21	14	17	11	13	12	15
5b	09	12	07	09	07	09	04	10
5c	12	15	10	11	10	13	08	11
5d	19	22	15	19	16	18	17	21
5e	08	11	11	14	04	08	03	07
Ofloxacin	23	25	21	26	-	-	-	-
Fluconazole	-	-	-	-	21	23	22	25

4. Discussion :

we have synthesized some N-substituted derivatives of 1,3,4-oxadiazole with the combination of substituted acid hydrazide and aryl isothiocynates in the presence of KI/I$_2$ in alkaline solution of NaOH. As a result, the synthesis of oxadiazole derivatives has been documented utilizing various techniques. Despite this, several other approaches have been outlined. In contrast to earlier published cyclization methods for the 1,3,4-oxadiazole ring, the experimental conditions described in this article made it possible for the heterocyclization process to be carried out with a relatively straightforward operation. This was made possible as a result of the fact that the heterocyclization was carried out under the conditions described in this article.

Consequently, a high yield (ranging from 60 to 95%) was accomplished in an appreciably shorter period of time. TLC analysis was carried out to determine the purity level held by the synthesized compounds. FT-IR, ^{1}H-NMR, and mass spectrometers were used to validate the synthesized compounds' structural properties. This was done to validate the compounds. The characteristic 1450-1480 cm^{-1} (C=N stretching) peaks and medium strong band at 1320-1340 cm^{-1} were identified in each IR spectra depicted 1,3,4-oxadiazole ring.

The results obtained by this study show that most of the synthesized compounds showed significant antimicrobial activity against S. aureus and E.coli bacteria in 50 and 100µg/ml concentrations with respect to the standard drug ofloxacin. Among the synthesized compounds **5a** and **5d** showed significant antibacterial activity against both bacteria and rest of the compounds showed good to moderate activity. Similarly, These were tested against A. niger and C. albicans fungi in 50 and 100µg/ml concentration with respect to the standard drug fluconazole. Among which **5d** showed significant antifungal activity while rest compounds revealed good to moderate antifungal activity. The results are represented in the Fig 2 and Fig 3 for antibacterial and antifungal activity respectively. The reason for high antimicrobial activity of these compounds could be the presence of electron donor group –Cl.

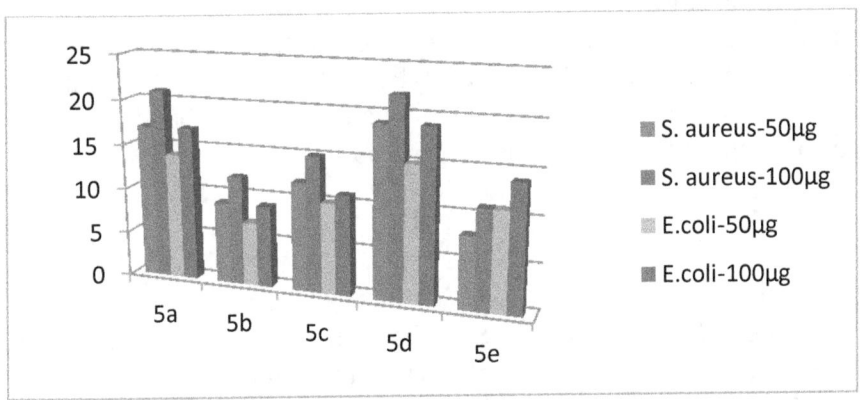

Fig: 2- Antibacterial activity of compound 5(a-e)

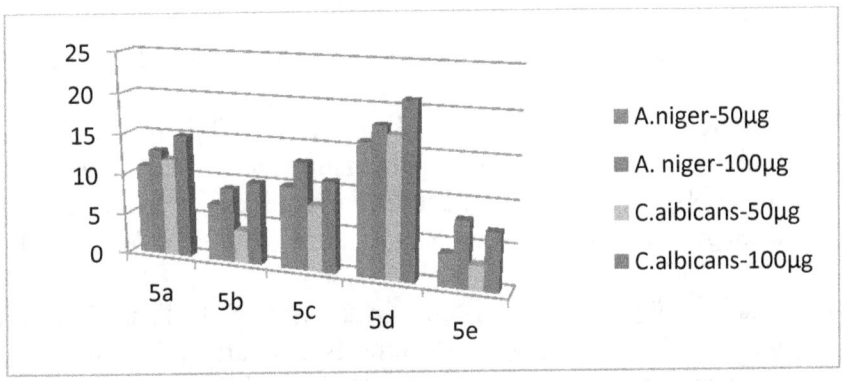

Fig: 3- Antifungal activity of compound 5(a-e)

5. Conclusion :

In the present work The structure of synthesized compounds was demonstrated by ^{1}HNMR, IR , ^{13}CNMR , Mass spectroscopy and elemental analysis. These compounds of series 5(a-e) were evaluated for antibacterial and antifungal activity against bacteria E. coli & S. aureus and fungus C. albicans & A. niger. Among newly formed compounds 5-(4-nitrophenyl)-*N*-phenyl-1,3,4-oxadiazol-2-amine **(5a)** and *N*-(4-chlorophenyl)-5-(4-nitrophenyl)-1,3,4-oxadiazol-2-amine **(5d)** revealed significant antimicrobial activity. The present research can support the new researchers for further characterization and evaluation of different derivatives of 1,3,4-oxadiazole compounds for the development of new useful antimicrobial agents.

6. Acknowledgement :

Authors are grateful to CSIR-HRDG, Delhi, for funding this project under the CSIR-UGC NET-JRF scheme. The authors also thank SAIF, Panjab University, and Chandigarh for performing spectral analysis.

References :

1) Eicher T, Hauptmann S. The chemistry of heterocycles: structure, reactions, syntheses, and applications. Thieme Medical Publishers Inc: New York; 1995.
2) Dua R, Shrivastava S, Sonwane SK, Srivastava SK. Pharmacological significance of synthetic heterocycles scaffolds a review. Advances in Biological Res. 2011; 59(3):120- 44.
3) Patel R. V., Patel P. K., Kumari P., Rajani D. P. & Chikjalia K.H., Synthesis of benzimidazolyl-1,3,4-oxadiazol-2ylthio-N-phenyl (benzothiazole) acetamides as antibacterial, antifungal and antituberculosis agents, Eur J. Med. Chem, 2012; 53:41.
4) Ş.Güniz Küçükgüzel, E.Elçin Oruç, Sevim Rollas, Fikrettin Şahin, Ahmet Özbek, Synthesis, characterization and biological activity of novel 4-thiazolidinediones, 1,3,4-oxadiazoles and some related compounds, Eur. J. Med. Chem., 2002; 37 (3):197-206.

5) Patel R., Kumari P. & Chikhalia K., Synthesis of coumarin-based 1,3,4-oxadiazol-2ylthio-N-phenyl/benzothiazole acetamides as antimicrobial and antituberculosis agents, Medicinal Chemistry, 2013 ;22:195–210
6) Omar F.A., Mahfouz N.M., Rahman M.A., Design, synthesis and anti-inflammatory activity of some 1,3,4-oxadiazole derivatives, Eur. J. Med. Chem. 1996; 31:819-25
7) T. Ozyazici, E.E. Gurdal, D. Orak et al., synthesis, anti-inflammatory activity and molecular docking studies of Some novel Mannich bases of the 1,3,4-oxadiazole-2(3H)-thione scaffold, Arch. Pharm., e2000061. 2020
8) Li S., Wang Z., Wei Y., Wu C., Gao S., Jiang H., Zhao X., Yan H., & Wang X., Antimicrobial activity of a ferrocene-substituted carborane derivative targeting multidrug-resistant infection, Biomaterials,;34:902. 2013
9) Priscilla P, Khadera AMA, Balakrishna K, Vinaya C.Synthesis of new 5-naphthyl substituted 1,3,4-oxadiazole derivatives and their antioxidant activity.Der Pharma Chemica. 2013; 5(2):24-3.
10) GB Patent 902388 - Improvements in and relating to new derivatives of 1,3,4oxadiazole and process for preparing same.
11) Savarino A. A historical sketch of the discovery and development of HIV-1 integrase inhibitors.Expert Opin. Investig. Drugs. 2006; 15: 1507-22.
12) Chiang J, Hermodsson G, Oie S. The effect of alpha 1-acid glycoprotein on the pharmacological activity of alpha 1-adrenergic antagonists in rabbit aortic strips. The Journal of pharmacy and pharmacology.1991; 43 (8): 540-7.
13) James ND, Growcott JW. Zibotentan. Drugs Future.2009; 34l: 624-33.
14) G.D. Hurd and G.D. Kelso; Boronic acid and ester inhibitors of thrombin; *J. Am. Chem. Soc.*; 62, 2184(1942).
15) S.S. Kumari, R.K.M Rao and N.V.S. Rao; Synthesis of Gamma-Pyrono (6, 7) -benz- (1, 2) Isoxazole and Their Spectral Characteristic; *Ind. J. Chem.* 11, 541(1973).
16) B.M. Bhawal, Ph.D. Thesis; Studies in the chemistry of benzisoxazoles, Marathawada Univ. (1977).

17) P.N. Wadodkar and M.G. Marthey; *Ind. J. Chem.* 10, 145(1972).
18) R.B. Golan & N. Paster, Mycotoxins in fruit and vegetables, 1st Ed., *Acad. Press Elsevier*(2008)
19) F.S. Stewart; Bigger's Handbook of Bacteriology, 9th Ed. Baltimore; *William and Wilkins Co.* (1968).
20) B.L.Oser, "Hawks, Physiological Chem." 4th Ed., *Tata McGraw Hill,* New Delhi(1978).
21) Sridhar SR, Rajagopal RV, Rajavel R, Masilamani S, Narasimhan S., Antifungal activity of some essential oils. J. Agric. Food. Chem., 2003, 51: 7596-7599.
22) H. munch, J.S. Hansen, M. Pittelkow, J.B. Christensen and U. Boas, A new efficient synthesis of isothiocynates from amines using di-tert-butyl-dicarbonate, Tetr. Lett., 2008.
23) Zaidan MRS, Rain AN, Badrul AR, Adlin A, Norazah A, Zakiah I. In vitro screening of five local medicinal plants for antibacterial activity using the disc diffusion method. Trop. Biomed., 2005, 22: 165-170.

[1]**Principal,**
Govt. College Tijara, Alwar
[2]**Research Scholar,**
Raj Rishi Bhartrihari Matsya University, Alwar
[1]sharma_ak002@yahoo.com
[2]nidhi.agarwal1210@gmail.com

5. Nature's Insurgence and Solidarity : Milton's Travel to Corporal and Supernatural Environment in *Paradise Lost*

Dr. Nipun Chaudhary

Abstract

John Milton is well known English classic poet who has been widely applauded for this book. This work endeavors to explore the components of Paradise Lost the mysteries of the sky in John Milton's epic poem and talks about the significance of *Paradise Lost*. In this epic poem John Milton explored about the story of Adam and Eve, those who ate that fruit from the Forbidden tree that brought trouble and death to men. John Milton has portrayed Satan as a hero of this poem and he also talks about the human kind of Satan. As a reader we think that Satan is a villain of the poem Paradise Lost but Satan is not a villain. The author of *Paradise Lost* is a third person which means he is not a part of the character in the poem.

Keywords : *Environment, Nature, Mythology, Solidarity, Ubiquity*

Introduction

After the underlying bewilderment of thick language and various implications, another peruses of Milton's *Paradise Lost* before long experiences a progressively immovable issue: the elucidation of God's job in the heroic sonnet. This is an issue on the grounds that the God of *Paradise Lost* isn't a persuading portrayal regarding supremacy, ubiquity, omniscience, outright goodness, equity, or shrewdness.

Inside *Paradise Lost,* notwithstanding, the trio isn't as noticeable all things considered in customary Christian regulation. Truth be told, Milton all things considered overlooked the presence of God in distinction to his scriptural heroic poem. Religious and scholar works of Milton's works minimize the job of the presence of God while talking about God. Furthermore, child, ruler and subject, leader and subordinate, however, Jesus unmistakably communicates a free will power once he offers for execution and undergo the grand keeper in the battle in contrast to Satan.

Pollution and climate change

The activities generate Jesus a legend over the whole of universe and in paradise, which is enough to generate to make a persuading contention that Jesus is the brave hero of the heroic. Jesus, at that point, is a main character in, however in his individual privilege and not as an additional portrayal of God in obvious structure. Besides, the various communications among God what's more, Jesus style it even more dire for Milton's to separate the portrayal of God and to comprehend its methods of activity. As per the topic of ethical discussion, God is not normal for whatever additional character may highlight in a heroic ballad. In the Christian perspective recognizable to Milton, God occur outdoor of the existence, and thus takes synchronous contact to wholly occasions and all places. God likewise has each capacity to change physical reality but then stay unaltered. At long last, God exemplifies without a doubt the decency that all things endeavor to turn into. Can an inescapable, all-powerful, and completely great being be an anecdotal character in any typical feeling of the word? Such a character, on the off chance that it was, would not be helpless to the human imperfections and irregularities that would style maximum anecdotal aspect fascinating.

Human characters enthrall my enthusiasm with their internal intentions and shrouded contemplations. I need to comprehend their lives and, over those lives, comprehend my individual. The most convincing charms are the utmost relatable. In any case, to express an enthusiasm for the thought processes and thinking of God is added issue. This intrigue addresses a powerful being of a nature entirely unexpected from what I understand. With God, I have a secret in the feeling of an entity obscure.

In this manner, a separation rises among God in evident structure and God as a character in the sonnet who does not seem on the way to consume each planetary, embrace each power, and typify goodness? This separation is vital for the incorporation of God in *Paradise Lost* as a total character also, presents an equivocalness in the sonnet about the genuine idea of God that shows up all through the fanciful portrayals, character improvement, and talks of God. I drive to think about this equivocalness in the portrayal of God by way of a puzzle of the sky.

There were two main great influences of Milton that worked upon the poetic career; they were the spirit of the Renaissance and the Reformation. The renaissance is the name given to the improving of Greek and Latin learning that happened in Europe during the 15th and 16th centuries. The renaissance brought with it not solely Associate in Nursing exaggerated interest in Greek and Latin literatures however additionally an increased interest in Arts and a larger enthusiasm forever as an entire and a stronger appreciation of lovely, bright and joyous in life. Milton's views developed from his awfully in depth reading in addition as travel and skill from his student days of the 1620s during the civil war in England. He died in 1674.

Characterizing of God

Paradise Lost ensures not just merely Christianity understanding; but it also typifies and comprises Christianity practice. What more is that Milton God existed first before anyone had, preeminent as an anecdotal appeal in an account, and to peruse the heroic ballad is to meet and communicate with God in its most genuine structure. Its own account has been reported in *Paradise Lost,* as opposed to suggesting and toying with stories from the legends of different writings what's more, conventions, that the intellectual effort first goes too far from backhanded to coordinate cases about the idea of God.

As an aspect in this sonnet, God uncovers quite a bit of its humor over revealed activity. In Milton's religious philosophy he brings up Gods work as a character about honesty and ethical Soundness. In Book V, the heavenly attendant Raphael grants to Adam a perfect representation among Satan's discourse to the lieutenant rogues and God's discourse to Jesus as separate side gets ready for the conflict in paradise. Raphael depicts God as

"th' Eternal eye, whose sight discerns/ Abstruse thoughts" (PL V.711- 712)

The change from the plural structure to the solitary structure eye underscores that the aid structure speaks to, not just the two earthly eyes of a person, yet additionally understanding and foreknowledge as a fundamental, intelligible nature of the pioneer. The peruse gets God's control through his capacity as president of the super powers to deliberately plot for success in the fighting contrast to Satan.

After the change of numerous structures to the solitary eye underscores that the substantial structure speaks to, not just the two bodily eyes of a person, yet in addition knowledge and premonition as a basic, intelligible nature of the pioneer. The peruse gets God's control over his capacity as president of the grand powers to deliberately plan for triumph in the conflict counter to Satan. As key organizer, God has greatness in Secrets of the Sky beating Satan's powers, not for the basic motive that paradise is more grounded than heck, be that as it may, for the more profound cause that the shrewdness of God's way outflanks the traps of Satan's shrewdness. "Through eye and vision, Milton builds up a metonymy among vision and light when Raphael relates that, investigating the developments of Satan's powers, God,-from forth his holy Mount/ And from within the golden Lamps that burned/ Nightly before him, saw without the light/ Rebellion rising." (PL V.712-715)

"The detail of light in *the golden Lamps that burne* underpins the possibility that God's actual wellspring of intensity is dynamic dream, not corporal quality. The legends of the *holy Mount* and *the golden Lamps* make a similar construction amongst the Christian folklore of Mount Sinai and the Menorah in the Temple of Jerusalem from one viewpoint, furthermore, the great folklore of Mount Olympus. This epic similarly more shows that Milton is depicting God in shadow rather than God in complete.—what multitudes/ Were banded to oppose his high Decree;/ And smiling to his only Son thus said" (PL V.716-718).

A considerable lot of God's accounted for activities are absurd on the off chance that I attempt to comprehend them as far as the genuine idea of God's being. As an abstract development, they just bode well as far as God's job as a character in the sonnet, which for this situation is the job of the president in charge of key arranging

and military activity. For example, Raphael discloses to Adam in the twelfth book the triumph of God finished Satan at last decision, God likewise shows up in the character of a victorious military leader. "Raphael describe,-all Nations shall be blest. / Then to the Heav'n of Heavens he shall ascend" (PL X.450-451).
This representation of Satan is very hazardous on the off chance that I comprehend it at a shallow dimension, without representing God's job as a character in the ballad.

Conclusion

Words now and then catch recollections. They convey successful an encounter that partakes passed. A great essayist in this method of composing will look to make the side an unmistakable impression of the world that encompasses it. Keats and other Romantic writers would build up a sort of verse that communicates the character of a minute as it sneaks past the psyche. This authenticity gets to rule current verse and I acknowledge lyrics to the extent that they profoundly light up a thing in my life. However, Milton kept in touch with various thoughts in an alternate time. Milton world later 1660 stood not bleeding insurgence in contrast to an English ruler, the destiny of fights among monarchists and democrats. The most astounding point of Milton's verse is to offer admittance to the planetary involved by God and the supernatural request of the corporal world right away accessible to detect involvement. This task revives the epic ballad when the political courses of action that were the significant subject of the day have later stretched time ago blurred into the lack of definition of collections and documents.

References
Achinstein, Sharon(1999). Milton and the Fit Reader. *British Literature 1640-1789: A Critical Reader*. Ed. Robert De Maria. Malden: Blackwell Publishing.
Ayer, AJ(1952). *Language, Truth and Logic*. Dover Publications.

Carnes, Valerie(1970). "Time and Language in Milton's Paradise Lost". *English Literary History.* 37(4). Web. 23 Jan. 2018.
Milton, John(1825) *A Treatise on Christian Doctrine*, trans. Charles A. Sumner, Cambridge University Press.
Milton, John(2013). *Paradise Lost.* Ed. Alastair Fowler. Rutledge. (All references from this edition)
Vercingetorix (2018). *Encyclopedia Britannica.* Encyclopedia Britannica Inc.

Associate Professor in English
School of Liberal Arts and Humanities
Chandigarh University
SAS Nagar, Punjab, India
email : **nipun085@gmail.com**

6. Mycoremediation : A Sustainable Clean Technology for Better Tomorrow

Dr. Geetanjli*, Simranjot, Kiranjeet Kaur

Abstract

Environmental pollution is a serious issue of global concern. Presently, industrial and environmental biotechnology is mainly focussing on the development of "Sustainable clean technologies". The emphasis is laid on maximum production, reduced waste generation, treatment and conversion of waste in some useful form for achieving Sustainable development goals. Mushrooms have been known to the mankind since ancient times. Apart from their nutritional and medicinal value, mushrooms hold a tremendous potential to be explored as an efficient tool for bioremediation. Fungi have been proven to be a cheap, effective and environmentally sound way for removing a wide array of contaminants from damaged environments or waste water. Mushrooms have a remarkable ability to degrade different types of pollutants. Mycoremediation, thus represents a biological tool to degrade, transform or immobilize environmental contaminants with the help of mushrooms and other fungi. This technique relies on the efficient enzymes, produced by fungi for the degradation of various types of substrate and pollutants. Mushrooms use different methods such as Biodegradation, Biosorption and Bioconversion for removal of contaminants from polluted spots and provide cleaner environment. This article reviews the achievement and current status of mycoremediation technology for the suitable and safe treatment of waste. This review is also focused on environment safety aspects of mushroom cultivation on waste. Mycoremediation holds a great potential to clean environment for a better tomorrow. However, certain issues concerned with the technique also need proper attention to achieve the maximum benefit of this eco-friendly technique with sustainable approach.

Keywords : Mycoremediation, Enzymatic Degradation, Bioconversion, Biodegradation, Biosorption.

Pollution and climate change

Introduction :
Environment safety is the major issue of global concern. However, polluted air, soil and water are major obstacle in achieving the targets of cleaner environment. In developing countries, pollution accounts for the loss of 5% gross domestic product (GDP), consequently leading to 16% of the global deaths, and 25% of the most polluted regions .Sustainable Development Goals (SDGs) include food security and availability of clean drinking water for all..According to the World Health Organization report, 2.2 billion people do not have access to safe drinking water, while 144 million people are compelled to use contaminated water . It is anticipated that by 2025, about fifty percent of the world population will be living in the water-stressed regions. In the last decade of 20th century, it was estimated that 22 million hectares of land were polluted, based on the current rate of urbanization and industrialization; this number is expected to increase further. Apart from adverse effect of pollution effect on public health and the economy, it can also threaten food security, drinking water availability, and biodiversity. So, it becomes imperative to develop strategies that can combat the soil and water pollution due to toxic chemicals and heavy metals. Currently, there are various physical and chemical methods available for the removal and degradation of various recalcitrant and harmful chemicals from soil and water. However, these methods are expensive, produce toxic byproducts, and ineffective for low concentrated but highly toxic chemicals. A safe strategy needs to be developed, which can overcome these limitations and provide in situ remediation of the pollutants. The biological process of remediation display features like economic viability (Ayangbenro 2017) and repeated use of biomass, selective metal binding, effective desorption and recycling of desorbents. Different microorganisms like algae, bacteria, fungi, yeast have been employed to carry out biosorption. As per a report, by 2023 fast-growing bioremediation market is estimated to reach $186.3 billion, with an annual increase of 15.4%. Mycoremediation can be an economical, eco-friendly, and effective strategy to combat the ever-increasing problem of soil and water pollution.

Mushrooms have fructifications, growing out of a mass of mycelium. These are considered as one of the favorite delicacy in many parts of the world. The consumption of edible mushrooms is increasing due to a good content of proteins and trace minerals. Mushrooms have also been reported as nutraceuticals having anti-oxidant, immune stimulatory, anti-cancer, anti-inflammatory and anti-diabetic therapeutic properties .

Apart from this, mushrooms can also be employed for decontamination of the polluted environment. Fungi are considered as Nature's scavengers, due to their ability to break down complex plant and woody material into simpler form. Just like they break down complex carbon based plant cell structures, like cellulose and lignin, saprophytic fungi use their digestive enzymes to break down chemicals like hydrocarbons and pesticides. Fungi can break down larger hydrocarbon chains into smaller pieces, allowing for microorganisms and plants to get to work. Fungi can also extract and hyper accumulate heavy metals, concentrating them in the fruiting body of the mushrooms. Mushrooms belonging to the genera including *Agaricus, Boletus, Armillaria, Polyporus, Russula, Pleurotus, Termitomyces* (Fig.1)have been investigated by some researchers for the uptake of heavy metals (Raj et al. 2011). High accumulation potential and shorter life span are some of the advantages of using mushrooms as biosorbents .The degradation can be augmented by adding carbon sources such as sawdust, straw and corn cob at polluted sites (Rhodes,2014) Mushrooms can build up heavy metals in high concentrations in their bodies above maximum permissible concentrations (Kalac and Svoboda 2000) and can act as an effective biosorption tool (Das 2005).

Fig.1: Some of Mushroom species used in Bioremediation

Mycoremedial Potential of Mushrooms :

Fungi are considered as powerful planetary healers and disaster responders. From layering mushrooms spawn with contaminated soil and wood chips, to installing mycofilters to trap contaminants in runoff and polluted water, there are many different ways to explore mushrooms to earth repair. Grassroots mycoremediators can install mushroom beds to deal with contaminated soil, decrease erosion, and help recharge and heal damaged land. Mushrooms come to the rescue in making our environments free from harmful ,toxic pollutants. Mycoremediation can be effective against a variety of pollutants coming from households,agriculture fields, pharmaceutical and dyeing industries etc (Fig. 2).

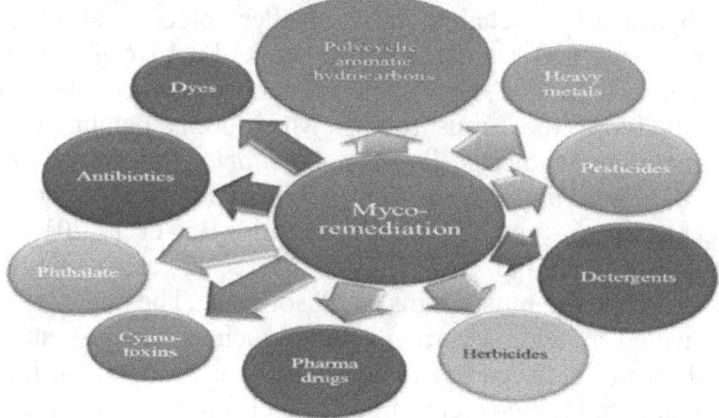

Fig.2: Various contaminants for which Mycoremediation can be used

In the past few years, increasing concentration of contaminants have become major eye sore for environmentalists all over the world (Duarte ,2018). Among these contaminants, the anthropogenic chemicals, endocrine-disrupting chemicals (EDCs) and pharmaceutical-personal care products (PPCPs) are relevant due to their biological effects on non target organisms; in particular, EDCs simulate or antagonize the endogenous hormone effects and are toxic to organisms also at very low concentrations. Estrone, 17β-estradiol, 17α-ethinylestradiol, bisphenol A and triclosan are the most detected and studied in soil. EDCs and PPCPs mainly enter the soil environment via irrigation with contaminated waste water(Ying and Kookana ,2005; Dodgen etal,2014) .

Polycyclic aromatic hydrocarbons (PAHs), derived from the incomplete combustion of organic materials, are molecules with multiple carbon rings, Their origin can be natural (e.g. open burning, natural losses of petroleum and volcanic activities) and predominantly anthropogenic (e.g. residential heating, , carbon black, coal gasification, petroleum refineries etc.). Polycyclic aromatic hydrocarbons contribute 13% to soil contamination as these have a tendency to bound to soil particles and to remain absorbed (Abdel-Shafy and Mansour, 2016). Both ligninolytic and non-ligninolytic fungi are able to degrade PAHs by means of the extracellular lignin-degrading enzymatic system, which contribute to the first attack on PAHs, and of the P450monooxygenase (Aranda,2016).

One of the first studies, about PAH degradation in soil microcosm, was carried out with *P. chrysosporium* and *T. harzianum*, grown on wheat straw and then inoculated in naphthalene-contaminated soil. The biodegradation behaviour was monitored by means of naphthalene concentration measurement, CO_2 evolution as well as phytotoxicity tests [80]. Apart from the model *P. chrysosporium*, various species of *Aspergillus, Penicillium, Rhizopus, Fusarium, Cladosporium* and *Trichoderma* are capable of PAHs degradation(Aydin,2017).

Significant Role of Mycoremediation:
Mycoremediation is comparatively economically cheaper method. Apart from decontamination of water and soil, it also helps to restore soil health as it is devoid of harmful chemicals. Mycoremediation also utilizes non-invasive species. Robust growth of fungus, vast hyphae network, production of versatile extracellular ligninolytic enzymes, high surface area to volume makes it an ideal choice to achieve cleaner environment. Various species of mushroom such as *Pleurotus, Agaricus, Coprinus* etc. have been reported to decontaminate soil of various contaminants (Fig.3). *Pestalotiopsis microspora*, the plastic eating mushroom consumes polyurethane, the key ingredient in plastic products, and converts it to organic matter (Fig. 4). Mycoremediation out performs other modes of bioremediation because of its eco-friendly, sustainable nature. Mycoremediation can even be used for fire management with the encapsulation method. For this process fungal spores coated with agarose in a pellet form are used. The pellet is

introduced to a substrate in the burnt forest, thus, breaking down the toxins in the environment and stimulating the growth. [Rhoades, 2014].

Fig.3 : Oyster mushrooms produced on oil contaminated soil .PC Fungi Perfecti (www.fungi.com)

Fig. 4: *Pestalotiopsis microspora* **, the plastic eating mushroom**

In ancient times, European folklore believed that the fairy rings, the circular growth of mushrooms on a grass lawn, were formed during gathering and dancing of fairies and elves. Today, it is well understood that fairy rings are formed as a natural phenomenon by the growth of mushrooms. The Fairy Rings, a mycoremediation project at Newtown Creek purposes to transform the water of toxins in a sculptural form. Oyster mushrooms are used in the Fairy Rings project for their ability to neutralize hydrocarbons present in petroleum products such as oil, pesticides, and herbicides and as a hyper accumulator to absorb heavy metals, common toxins found in urban environments (Fig.5). The sculptural form is an alliance between the material structure and nature. Nature participates through the ongoing process of sprouting mushrooms, mycelium decomposing and cleansing the water of toxins.

Fig.5: Fairy ring Project for decontamination of water

In 2007, a cargo ship spilled 58,000 gallons of fuel along the San Francisco shoreline. Mats woven from human hair (resembling doormats), were used as sponges to soak up the spilled oil. These were then collected and layered with oyster mushrooms and straw. The mushrooms broke down the oil and after several weeks the resulting soil was good enough to be used to for roadside landscaping.

Mechanism of Mycoremediation :

Mycoremediation can be both enzymatic as well as non-enzymatic (Fig.6) . Different methods used by mushrooms to decontaminate polluted spots and stimulate the environment include - (i) Biodegradation (ii) Biosorption (iii) Bioconversion.

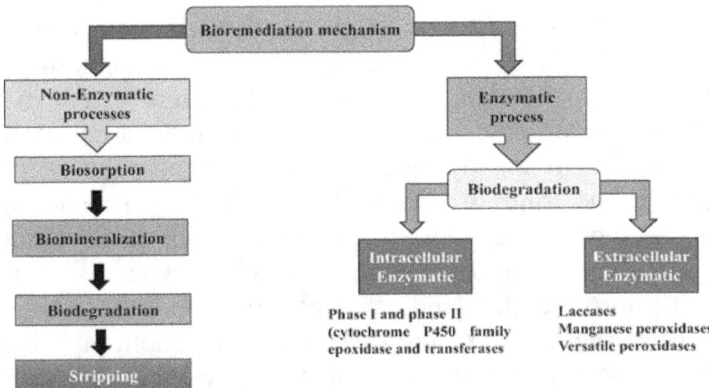

Fig.6: Mechanism of Mycoremediation

Biodegradation

The term 'Biodegradation' is used to describe the ultimate degradation and recycling of complex molecule to its mineral constituents. It is the process which leads to complete mineralization of the starting compound to simpler ones like CO_2, H_2O, NO_3 and other inorganic compounds by living organisms. The most suitable fungi to be used in soil remediation are basidiomycetes and, in particular, the ecological groups of saprotrophic and biotrophic fungi

Table 1: Mushrooms used in biodegradation:

Mushroom species	Type of Pollutant	Reference
Coriolus versicolor	PAH	Jang et al. (2009)
Lentinula edodes	2,4-dichlorophenol	Tsujiyama et al. (2013)
Pleurotus ostreatus	Oxo-Biodegradable plastic	da Luz et al. (2013)
Pleurotus pulmonarius	Crude oil; Radioactive cellulosic-based waste	Olusola and Anslem (2010); Eskander et al. (2012)
Schizophyllum commune and Polyporous sp.	Malachite green	Rajput et al. (2011)

Biosorption:

Biosorption is a process based on the sorption of metallic ions/pollutants/xenobiotics from effluent by live or dried biomass. Biosorption is considered as an alternative to the remediation of industrial effluents as well as the recovery of metals present in effluent. Cd, Cr, Hg, Pb, Cu, Zn and As are the most common heavy metals found in soil. Fungi in particular basidiomycetes members like mushrooms, are potential heavy metal accumulators, can uptake heavy metals from soil by means of their mycelia and accumulate the latter in the fruiting bodies(Raj et al ,2011). Biosorption is a process based on the sorption of metallic ions/pollutants/xenobiotics from effluent by live or dried biomass which often exhibits a marked tolerance towards metals and other adverse conditions (Gavrilescu (2004). Biosorbents can be prepared from mushroom mycelium and spent mushroom compost. A number of mushroom species have been reported to be used as biosorption of various heavy metals (Table 2).

Table 2: Mushrooms used in Biosorption

Mushroom species	Heavy Metal/Pollutant	Reference
Agaricus bisporus, Calocybe indica, Pleurotus platypus,	Copper, Zinc, Iron, Cadmium, Lead, Nickle	Lamrood and Ralegankar (2013)
Agaricus bisporus, Lactarius piperatus	Cadmium (II) ions	Nagy et al. (2014)
Flammulina velutipes	Copper	Luo et al. (2013)
Fomes fasciatus	Copper (II)	Sutherland and Venkobachar (2013)
Pleurotus ostreatus	Cadmium	Tay et al. (2011)
Pleurotus sajor-caju	Heavy metal Zn	Jibran and Milsee Mol (2011)
Pleurotus tuber-regium	Heavy metals	Oyetayo et al. (2012)

Bioconversion:

Presently, conversion of industrial or agro-industrial sludge into some other useful forms is going on. Any lignino-cellulosic waste, generated by industries, can be used for cultivation of mushroom which can be further use as a product. Mushroom species cultivated on industrial and agro-industrial wastes are given in Table 3. The choice of the substrate for the cultivation of mushroom is generally determined by the regional availability of the material.

Table 3: Bioconversion of agro-industrial waste in to mushrooms

Mushroom species	Reference	Type of agro/ industrial waste
Lentinula edodes	Gaitán-Hernández et al (2006); Brienzo et al (2007)	Vineyard pruning, barley straw, and wheat straw; Eucalyptus waste
Lentinula tigrinus	Lechner and Papinutti (2006)	Wheat straw
Pleurotus citrinopileatus, Pleurotus florida	Kulshreshtha et al (2010, 2013)	Handmade paper and cardboard industrial waste
Pleurotus tuber-regium	Jonathan et al. (2008), Kuforiji and Fasidi (2008)	Cotton waste, sawdust of Khaya ivorensis and rice straw
Pleurotus ostreatus	Akinyele et al (2012)	Handmade paper and cardboard industrial waste
Pleurotus sp.	Kuforiji and Fasidi (2009)	Cotton waste, rice straw, cocoyam peels and saw dusts of Mansonia altissima, Boscia angustifolia and Khaya ivorensis
Volvariella volvacea	Belewu and Belewu (2005); Akinyele et al. (2011)	Banana leaves, Agro-industrial residues such as cassava, sugar beet pulp, wheat bran and apple pomase

Factors Affecting Mycoremediation :

In general, various physical and chemical characteristics of soil such as pH, temperature, water content and redox potential etc. exhibit significant impact on the microbial growth and consequently on the success of a bioremediation process.

In particular, the biodegradation activity of the microorganisms depends on macro- and micronutrient availability in soil and on the presence of any other factor that influence the microbial metabolism, such as the contaminant type and concentration, and their bioavailability, toxicity and mobility.

A proper amount of nutrients for microbial growth is usually present in soil; nevertheless, nutrients can also be added in a functional form which serves as an electron donor to stimulate bioremediation process (Adams etal, 2015). The biodegradation of a toxic compound mainly depends on the genetic characteristics of the microorganism, in particular on both the extracellular and intracellular enzymatic systems(Chanda et al,2016) . The contaminant concentration directly influences the microbial activity, a high concentration may produce a variety of toxic effects on the different microbial classes, whereas a low concentration could not be enough to activate the synthesis of degradation enzymes. Filamentous fungi, able to form extended mycelial network and to synthetize a lot of aspecific enzymes, generally show a higher resistance to high contaminant concentration than bacteria(Harms et al,2011). Low substrate specificity, the synthesis of degradative enzymes occurs even at low concentration of contaminans. The intracellular metabolic pathways involved in mycoremediation show remarkable similarities with those that regulate the secondary metabolism in fungi, in particular those of mycotoxin production (Fig.7) Filamentous fungi which produce mycotoxins (e.g. *Aspergillus and Penicillium spp.*) exhibit the ability to degrade a wide variety of pharmaceutical compounds (Agunbiade and Moodle,2014) among them the emerging pollutants EDCs. In effectively degraded by bacteria.

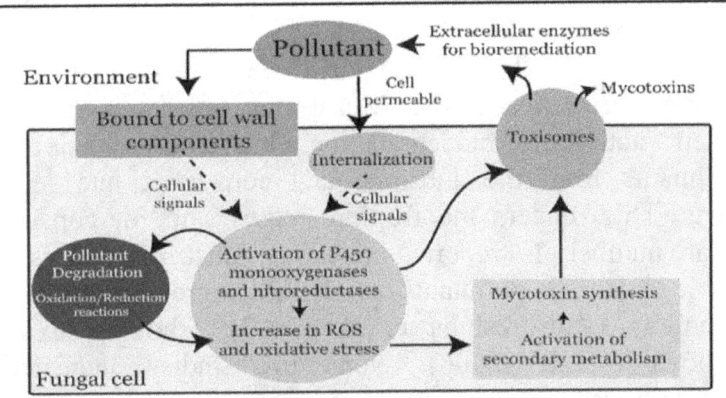

Fig.7: Metabolic pathways involved in Mycoremediation

The contaminant bioavailability is one of the most important factors that can be improved to optimize and accelerate the biodegradation; this fact has been demonstrated in the mycoremediation of aged PAH-contaminated soils(Leonardi,2007). The fungal ability to chemically modify or affect the contaminant bioavailability by means of biosurfactant production has been reported in different reviews(Shekhar,2015,Prakash,2017) . *Penicillium and Aspergillus* species have been reported to be biosurfactant producers . A wide range of microbial biosurfactant applications have been reported in the environmental protection field (e.g. enhancing oil recovery, controlling oil spills, biodegradation and detoxification of oil-contaminated soils) .

Challenges & Limitations in Mycoremediation :

Using fungi for bioremediation and earth repair requires a large amount of mycelium. Mycelium refers to the ultra-fine and dense network of branching thread-like white hyphae that is the vegetative part of the fungi. It is the mycelia that send out the enzymes that break down chemicals, and that also act as a filter, among many other things. In order to do effective bioremediation work and mycoremediation installations, a lot of mycelium is needed, and this can be challenge to most grassroots bioremediators. Mushroom cultivated on industrial wastes may possess toxicity/genotoxicity. Genotoxicity of mushrooms is influenced by genotoxicants that are present in waste used for their cultivation. Therefore, it is necessary to assess toxicity/genotoxicity of mushrooms if used for bioremediation purpose.

Biodegradation and biosorption influence toxicity level in the mushrooms fructifications. Mushroom possesses the suitable enzymatic activity for biodegradation which lead to the degradation of pollutants from the substrate and convert it into less toxic products. This renders the fruiting bodies safe for consumption. There are number of reports indicating that mushrooms are not only able to degrade pollutants but also able to reduce the toxicity or mutagenicity (Kulshreshtha et al. (2013); Choi et al. (2013); Malachová et al. (2006). Numerous studies indicated that mutagenicity reduction by mushrooms is primarily dependent on species. Kulshreshtha et al. (2011,2013) reported that when Two different species of *Pleurotus* were cultivated on industrial wastes and the mixture of wheat straw and industrial wastes under the same cultivation conditions *Pleurotus citrinopileatus* possessed genotoxicity in their fruiting bodies, while the same was lacking in *Pleurotus florida* .

Toxicity reduction also depends on the nature of substrate. Same fungi may have different capability in degradation of the different pollutants (Choi et al.,2013) due to the enzymes of mushrooms that are not only involved in degradation but also reducing the effects of toxic and genotoxic pollutants. Antimutagenic and antigenotoxic power of mushrooms have been demostrated by several researchers (Mendez-Espinoza et al. (2013); Taira et al. (2005); Mlinaric et al. (2004); Filipic et al (2002); Menoli et al. (2001). Thus, ii proved that besides having degradation power mushrooms can reduce the genotoxicants and toxic pollutants due to having antimutagenic and antigenotoxic power. These types of species of mushroom can be used for edible purposes and as feed for animals. This concept provides a natural guide to future research which should be focused on the need of research to degrade the pollutants in such a way that their disposal will not create another problem and fruiting bodies can be consumed safely. In contrast to this, absorption of pollutants by mushroom makes them unsuitable for consumption. A number of workers have reported high metal content and mutagenicity in the fruiting bodies of mushrooms growing on polluted substrate, naturally or artificially, which is due to the absorption process. More detailed information about the level of toxicity in these mushrooms is needed, ignorance of which may consequently lead to various health issues.

Discussion and Conclusion :
Mycoremediation can be an economical, eco-friendly, and effective strategy to combat the ever-increasing problem of soil and water pollution. In order to determine the best conditions for the process and toxicity in the fruiting bodies, it is extremely important to carry out feasibility study before starting a remediation project. Biodegradability, contaminant distribution, chemical reactivity of the contaminants, soil type , availability of oxygen and presence of inhibitory substances are the important parameters to define a contaminated site are: (Martín et al. (2004). The success of mycoremediation is governed by three important factors- availability of mushroom, accessibility of contaminants and a conductive environment. Therefore, the knowledge on the physiology and ecology of the biological species or consortia involved and the characteristics of the polluted sites are decisive factors to select an adequate mycoremediation protocol (Martín et al. 2004). Mushroom is a tremendous boon to the idea of using this for mycoremediation process as a real-world solution. The cultivation of edible mushroom on agricultural and industrial wastes may thus be a value added process capable of converting these discharges, which are otherwise considered to be wastes, into foods and feeds. It is extremely important to carry out feasibility study before starting a remediation project in order to determine the best conditions for the process and toxicity in the fruiting bodies Besides producing nutritious mushroom, it reduces genotoxicity and toxicity of mushroom species..

While decontamination can be a solution for some of the environmental problems created by toxic residues, it is clear that in order to avoid an even worse scenario than the current one, a profound change in how industries and population manage and perform their activities must happen, and this change must meet emerging sustainability criteria. Non-toxic residues that are generated as by-products from different industries become a problem as well, because they are produced in large amount and need appropriate and economical waste management. Many of these non-toxic wastes from food, animal feed and agricultural industries could be used as fungal substrate to produce fungal proteins or other useful metabolites, thus converting industrial by-products into valuable material. Degradation of many substrates, even toxic materials is an inherent characteristic of fungal metabolism, so incorporating fungi-based transformation processes to deal with the ever increasing amount of byproducts generated by our society needs, could very well be a way of saving the world with the help of fungi.

Mycoremediation of waste from the environment by mushroom has many advantages but at the same time it is a challenge for the researchers, environmentalists and engineers. Mycoremediation of wastes can be done in in situ and ex situ conditions. When it is carried out on site, it eliminates the need to transport the toxic materials to treatment sites. It is an environmentally friendly approach and needs only a small space, low cost, less skilled persons and can be applied easily in the field. In contrast to above, there are some disadvantages in applying this mycoremediation tool.

Mushrooms require time to adapt to the environment and cleanup wastes . Paul Stamets, a major protagonist of mycoremediation , has proposed that there should be "Mycological ResponseTeams", who would employ fungi to recycle and rebuild healthy soil in the area following any contamination incident like oil spill, chemical leak, radiation egress etc.(Stamets,2005).It has been suggested that edible mushrooms might be grown for the purposes of mycoremediation, and the prospects of whether they would be safe to eat afterwards have to be considered(Kulshreshtha et al, 2014). Naturally this depends on the exact nature of the pollutants, so that heavy metals are likely be a problem ,if they are absorbed and concentrated into the mushroom, while some organic soil contaminants might be decomposed without so imparting toxicity. In the latter case, the benefit is offered that land that is contaminated and unfit for agriculture could be both cleaned and made to yield a nutritious food crop. Mycoremediation holds a promising role as a sustainable clean technology for a, better tomorrow Mycoremediation through mushroom cultivation will alleviate two of the world's major problems i.e. waste accumulation and production of proteinaceous food simultaneously. Thus, there is a need for further research towards the exploitation of potential of mushroom as bioremediation tool for decontamination purposes and its safety aspects for consumption as product.

References :
Abdel-Shafy H.I., Mansour M.S.M.(2016) A review on polycyclic aromatic hydrocarbons: Source, environmental impact, effect on human health and remediation, **Egyptian Journal of Petroleum,** 25,107-123.
Adams G.O., Fufeyin P.T., Okoro ,S.E., Ehinomen I. (2015)Bioremediation, biostimulation and bioaugmention: A review. **International Journal of Environ Bioremed & Biodegrad.** 3(1),28-39.
Adebayo, A.O. (2013) Investigation on *Pleurotus ferulae* potential for the sorption of Pb (II) from aqueous solution. **Bull Chem Soc Ethiop,** 27, 25–34.
Adenipekun, C.O.(2008) Bioremediation of engine-oil polluted soil by *Pleurotus tuber-regium* Singer, a Nigerian white-rot fungus. **Afr J Biotechnol** ,7, 55–58.
Adenipekun, C.O., Ogunjobi A.A., Ogunseye, A.O. (2011) Management of polluted soils by a white-rot fungus: *Pleurotus pulmonarius*. **Assumption Univ J Technol**, 15, 57–61.
Adenipekun, C.O. and Lawal, R. (2012) Uses of mushrooms in bioremediation: A review. Biotechnol. Molec. **Biol. Rev.**, 7(3) 62-68.
Agunbiade, F.O., Moodle, B.(2014). Pharmaceuticals as emerging organic contaminants in Umgeni River water system, KwaZulu-Natal, South Africa. **Environmental Monitoring and Assessment**. 186,7273-7291.
Akinyele, B.J., Olaniyi, O.O., Arotupin, D.J., (2011) Bioconversion of selected agricultural wastes and associated enzymes by *Volvariella volvacea*: An edible mushroom. **Res J Microbiol.**,4,

63–70. doi: 10.3923/jm.2011.63.70. doi:10.3923/jm.2011.63.70.

Anand, P., Isar, J., Saran ,S., Saxena, R.K. (2006) Bioaccumulation of copper by *Trichoderma viride*. **Bioresour Technol.**, 97, 1018–1025.

Aranda, E. (2016)Promising approaches towards biotransformation of polycyclic aromatic hydrocarbons with Ascomycota fungi. **Current Opinion in Biotechnology.**,38 ,1-8.

Ayangbenro Babalola,(2017) A new strategy for heavy metal polluted environments: a review of microbial biosorbents. **Int J Environ Res Public Health**, 14,94.

Aydin S, Karacay HA, Shahi A, Gokce S, Ince B, Ince O. (2017) Aerobic and anaerobic fungal metabolism and Omics insights for increasing polycyclic aromatic hydrocarbons biodegradation. **Fungal Biology Reviews,**31,61-72.

Belewu, M.A. Belewu, K.Y.(2005) Cultivation of mushroom (*Volvariella volvacea*) on banana leaves. **African J Biotechnol.**;4,1401–1403.

Bhattacharya, S.,Das, A M. G, V. K, S. J, (2011) , Mycoremediation of Congo red dye by filamentous fungi, **Brazilian J. Microbiol**. 42,1526.

Bhattacharya, S., Das, A., Prashanthi, K. Palaniswamy, M. , Angayarkanni, J. (2014) Mycoremediation of Benzo[a]pyrene by Pleurotus ostreatus in the presence of heavy metals and mediators, 3 Biotech. 4,205–211. https://doi.org/10.1007/s13205-013-0148-y

Bhattacharya, S. et al. (2014) Mycoremediation of Benzo[a]pyrene by *Pleurotus ostreatus* in the presence of heavy metals and mediators. **Biotech**,4(2) , 205-211.

Brienzo M, Silva EM, Milagres AM.(2007) Degradation of eucalyptus waste components by *Lentinula* edodes strains detected by chemical and near-infrared spectroscopy methods. **Appl Biochem Biotechnol.**;4:37–50. doi: 10.1007/s12010-007-9209-1.

Chanda ,A., Gummadidala, P.M., Goma, O.M.(2016) Mycoremediation with mycotoxin producers: A critical perspective. **Applied Microbiology and Biotechnology**.100,17-29.

Choi YS, Long Y, Kim MJ, Kim JJ, Kim GH. Decolorization and degradation of synthetic dyes by Irpex lacteus KUC8958. J Environ Sci Health A Tox Hazard Subst Environ Eng. 2013;4:501–508.

Das, N.(2005) Heavy metals biosorption by mushrooms. **NPR.**,4, 454–459.
Da Luz, J.M.R., Paes ,S.A., Nunes, M.D., da Silva, M.C.S., Kasuya, M.C.M. (2013) Degradation of Oxo-Biodegradable Plastic by *Pleurotus ostreatus*. **PLoS ONE**.;4(8):69386. doi: 10.1371/journal.pone.0069386. doi:10.1371/journal.pone.0069386.
Dodgen L.K., Li j WX, Lu Z, Gan J.J. (2014)Transformation and removal pathways of four common PPCP/EDCs in soil. **Environmental Pollution**. 193,29-36.
Du, G. Pu, C. Shao, S. Cheng, J. Cai, L. Zhou, Y. Jia, X. Tian, (2015) Potential of extracellular enzymes from Trametes versicolor F21a in Microcystis spp. degradation, **Mater. Sci. Eng. C.** 48, 138–144.
Duarte R.M.B.O., Matos, J.T.V., Senesi, N.(2018) Organic pollutants in soils. In: Duarte AC, Cachada A, Rocha-Santos T, editors. Soil Pollution From Monitoring to Remediation. London, UK: Elsevier/Academic Press; pp. 103-126.
Dulay, R.M.R., Parungao, A.G., Kalaw, S.P., Reyes, R.G.(2012) Aseptic cultivation of *Coprinus comatus* (O. F. Mull.) Gray on various pulp and paper wastes. **Mycosphere**.,4,392–397. doi: 10.5943/mycosphere/3/3/10. doi:10.5943/mycosphere/3/3/10. 392.
Fawzy, E.M., Abdel-Motaa,l F.F., EL-zayat, S.A.(2017) Biosorption of heavy metals onto different eco-friendly substrates. **J Toxicol Environ Health Sci**, 9(5), 35–44.
Filipic, M., Umek, A., Mlinaric ,A.,(2002) Screening of Basidiomycete mushroom extracts for antigenotoxic and bio-antimutagenic activity. **Pharmazie**.;4:416–420.
Firdousi, S.A.(2017) Bioaccumulation and bio-absorptions of heavy metals by the mushroom from the soil. **J Med Chem Drug Discov**, 2(3), 25–33.
Gaitán-Hernández ,R., Esqueda, M., Sánchez, A., Beltrán-García, M., Mata,G.,
(2006) Bioconversion of agrowastes by *Lentinula edodes*: The high potential of viticulture residues. **Appl Microbiol Biot**.,4,432–439.
Gavrilescu, M. (2004) Removal of heavy metals from the environment by biosorption. **Eng Life Sci**.4,219–232.
Harms H, Schlosser D, Wick LY.(2011) Untapped potential:

Exploiting fungi in bioremediation of hazardous chemicals. Nature Reviews. **Microbiology.** 9,177-192.

Jakovljević, J., Vrvić, M.M. V.D. (2017) *Penicillium verrucosum* as promising candidate for bioremediation of environment contaminated with synthetic detergent at high concentration, **Appl. Biochem. Microbiol.** 53, 368–373.

Jang, K.Y., Cho S.M., Seok, S.J., Kong, W.S., Kim, G.H., Sung, J.M.(2009) Screening of biodegradable function of indigenous ligno-degrading mushroom using dyes. **Mycobiology**.4,53–61. doi: 10.4489/MYCO.2009.37.1.053. doi:10.4489/MYCO.2009.37.1.053.

Jibran, A.K., Milsee Mol, J.P.(2011) *Pleurotus sajor-caju* Protein: A potential biosorptive agent. **Adv Bio Tech**,4, 25–27.

Jonathan ,S.G., Fasidi, I.O., Ajayi, A.O., Adegeye, O. (2008)Biodegradation of Nigerian wood wastes by *Pleurotus tuber-regium* (Fries) Singer. **Bioresource Technol.** 4,807–811. doi: 10.1016/j.biortech.2007.01.005. doi:10.1016/j.biortech.2007.01.005.

Kahn I, Aftabb M, Shakir S, Ali M, Qauum S, Haleem KS, Touseef I, (2019) Mycoremdiation of Heavy metal (Cd and Cr) though indigenous metallo fungal isolates, **Environ. Monit. Assess**. 191. https://doi.org/10.1007/s10661-019-7769-5

Kapahi, M. and Sachdeva, S.(2017) Mycoremediation potential of *Pleurotus* species for heavy metals: a review, **Bioresources and Bioprocessing**, 4,32 .

Koul,B., Ahmad, W. and Singh, J. (2021)Mycoremediation: A novel approach for sustainable development, **Food Science, Technology and Nutrition**, 409-420,

Kuforiji, O.O., Fasidi , I.O.(2008) Enzyme activities of *Pleurotus tuber-regium* (Fries) Singer, cultivated on selected agricultural wastes. **Bioresource Technol.** 4, 4275–4278. doi: 10.1016/j.biortech.2007.08.053. doi:10.1016/j.biortech.2007.08.053.

Kulshreshtha ,S., Mathur, N., Bhatnagar, P., Jain, B.L.((2010) Bioremediation of industrial wastes through mushroom cultivation. **J Environ Biol.** 4, 441–444.

Kulshreshtha, S., Mathur, N., Bhatnagar, P. (2011)Pros and cons of *P.florida* cultivation for managing waste of handmade paper and cardboard industries. **IIOAB Journal**, spl.4,5–48.

Kulshreshtha, S., Mathur, N., Bhatnagar, P.(2013)

Mycoremediation of paper, pulp and cardboard industrial wastes and pollutants, In: Fungi as Bioremediators: Soil Biology. Goltapeh EM, Danesh YR, Varma A, editor. Springer Berlin, Heidelberg, pp. 77–116.

Kulshreshtha, S., Mathur, N., Bhatnagar, P.,(2013)Cultivation of *Pleurotus citrinopileatus* on handmade paper and cardboard industrial wastes. **Ind Crop Prod.**,4: 340–346. doi: 10.1016/j.indcrop.2012.04.053.

Kulshreshtha S, Mathur N, and Bhatnagar P,(2014)Mushroom as a product and their role in mycoremediation, **AMB Express**. 4,29. doi: 10.1186/s13568-014-0029-8.

Kumhomkul, T., Panich-pat, T.(2013) Lead accumulation in the straw mushroom, *Volvariella volvacea,* from lead contaminated rice straw and stubble. **Bull Environ Contam Toxicol.**,4,231–234.

Lamar, R.T. and White, R.B. (2001) Mycoremediation: commercial status and recent developments. In: V.S. Magar et al. (eds), **Proc.Sixth Int. Symp. on In Situ and On-Site Bioremediation,**6:263-278. San Diego, CA.

Lamrood ,P.Y., Ralegankar, S.D.(2013) Biosorption of Cu, Zn, Fe, Cd, Pb and Ni by non-treated biomass of some edible mushrooms. **Asian J Exp Biol Sci.**;4: 190–195.

Lechner B.E., Papinutti, V.L. (2006) Production of lignocellulosic enzymes during growth and fruiting of the edible fungus *Lentinus tigrinus* on wheat straw. **Process Biochem.**;4: 594–598.

Leonardi V., Sasek ,V., Petruccioli, M. D'.Annibale, A., Erbanova, P., Cajthaml ,T.(2007) Bioavailability modification and fungal biodegradation of PAHs in aged industrial soils. International Biodeterioration and Biodegradation.;60:165-170.

Luo, D., Yf X, Tan ,Z.L., Li, X.D.,(2013) Removal of Cu^{2+} ions from aqueous solution by the abandoned mushroom compost of *Flammulina velutipes.* **J Environ Biol.**;4:359–365.

Martín, M.C., González, B.A., Blanco, S.M.J. (2004) Biological treatments for contaminated soils: hydrocarbon contamination. Fungal applications in bioremediation treatment. **Rev Iberoam Micol.**4:103–120.

Mendez-Espinoza, C., Garcia-Nieto, E., Esquivel ,A.M., Gonzalez, M.M., Bautista, E.V., Ezquerro, C.C., Santacruz, L.J.,(. 2013)

Antigenotoxic potential of aqueous extracts from the chanterelle mushroom, *Cantharellus cibarius* (higher Basidiomycetes) on human mononuclear cell cultures. **Int J Med Mushrooms**,4,325–32
Menoli, R.C., Mantovani, M.S. Ribeiro, L.R., Speit, G., Jordão, B.Q.(2001) Antimutagenic effects of the mushroom *Agaricus blazei* Murrill extracts on V79 cells. **Mutat Res**.;4:5–13.
Mlinaric A, Kac J, Fatur T, Filipic M. Anti-genotoxic activity of the mushroom Lactarius vellereus extract in bacteria and in mammalian cells in vitro. **Pharmazie**. 2004;4:217–221
Nagy, B., Măicăneanu, A., Indolean, C., Mânzatu ,C., Silaghi-Dumitrescu, M.C.(2014), Linear and nonlinear regression analysis for heavy metals removal using *Agaricus bisporus* macrofungus,**Arabian Journal of Chemistry**, 148(S2).
Nagy, B., Măicăneanu, A., Indolean, C., Mânzatu ,C., (2014)Comparative study of Cd(II) biosorption on cultivated *Agaricus bisporus* and wild *Lactarius piperatus* based biocomposites. Linear and nonlinear equilibrium modelling and kinetics, **Journal of the Taiwan Institute of Chemical Engineers**, 45(3),921–929.
Olusola, S.A., Anslem, E.E.(2010) Bioremediation of a crude oil polluted soil with *Pleurotus pulmonarius* and *Glomus mosseae* using *Amaranthus hybridus* as a test plant. **J Bioremed Biodegrad**.;4,111. doi:10.4172/2155-6199.1000113.
Oyetayo, V.O., Adebayo, A.O., Ibileye, A.(2012) Assessment of the biosorption potential of heavy metals by Pleurotus tuber-regium. **Int J Advanced Biol Res**.;4:293–297.
Prakash ,V.(2017) Mycoremediation of environmental pollutants. **Int J Chem Tech Res**; 10(3),149–155.
Prasad, A.S.A., Varatharaju, G., Anushri, C., Dhivyasree, S.(2013) Biosorption of lead by *Pleurotus florida* and *Trichoderma viride*. **Br Biotechnol J**;3(1):66–78.
Raj, D.D., Mohan, B., Vidya Shetty, B.M. (2011)Mushrooms in the remediation of heavy metals from soil. **Int J Environ Pollut Control Manag** , ,3(1),89–101.
Rajput, Y., Shit, S., Shuklam, A. Shukla ,K..(2011) Biodegradation of malachite green by wild mushroom of Chhatisgarh. **J Exp Sci.**, 4,69–72.

Rhodes, C.J. (2014) ,Mycoremediation (bioremediation with fungi) –growing mushrooms to clean the earth, **Chemical Speciation & Bioavailability**,26(3), 196-198 .

Rodríguez Eugenio, N., Michae F., McLaughlin, l.,. Bazza, Z.,. Ronald Vargas, F , Kohlschmid, E., Oxana Perminova, F. Elisabetta Tagliati, , F. ,. Khan, A .. Pennock, L . (2018) Hidden Soil Pollution: A Reality ,Food and Agriculture Organization of the United Nations Edition, Design & Publication.

Salman, H.A., Ibrahim, M.I., Tarek , M.M., Abbas, H.S.(2014) Biosorption of heavy metals—a review. **J Chem Sci Technol**, 3(4), 74–102.

Shekhar, S., Sundaramanickam ,A., Balasubramanian, T.(2015) Biosurfactant producing microbes and their potential applications: A review. **Critical Reviews in Environmental Science and Technology.**;45 (14),1522-1554.

Singh, H.(2006), Mycoremediation: fungal bioremediation, Wiley,Hoboken.

Singh, M., Srivastava, P.K.. Verma,, P.C.,. Kharwar, R.N Singh, N. , Tripathi, R.D. , Soil fungi for mycoremediation of arsenic pollution in agriculture soils, **J. Appl. Microbiol.** 119 (2015) 1278–1290. https://doi.org/10.1111/jam.12948.

Sutherland C, Venkobachar C. (2013) Equilibrium modeling of Cu (II) biosorption onto untreated and treated forest macro-fungus *Fomes fasciatus*. **Inter J of Plant, Animal and Envi Sci** .4,193–203.

Stamets, P. (2005), Mycelium running: how mushrooms can help save the world. Ten Speed Press, Berkley.

Rani V. &Das N. (2011) Mechanism of Cd(II) adsorption by macrofungus *Pleurotus platypus*. **J Environ Sci** ,23:288–293.

Taira, K., Miyashita, Y., Okamoto, K., Arimoto, S., Takahashi, E., Negishi, T., Novel antimutagenic factors derived from the edible mushroom Agrocybe cylindracea. **Mutat Res**. 2005;4:115–123.

Tay CC, Liew HH, Yin CY, Abdul-Talib S, Surif S, Suhaimi AA, Yong SK.(2011) Biosorption of Cadmium ions using Pleurotus ostreatus: Growth kinetics, isotherm study and biosorption mechanism. **Korean J Chem Eng.**,4,825–830.

Tsujiyama S, Muraoka T, Takada N.(2013) Biodegradation of 2,4-

dichlorophenol by shiitake mushroom (Lentinula edodes) using vanillin as an activator. **Biotechnol Lett.**,4, 1079–1083. doi: 10.1007/s10529-013-1179-5.

Vaseem, H., Singh, V.K. Singh, M.P (2017) Heavy metal pollution due to coal washery effluent and its decontamination using a macrofungus, Pleurotus ostreatus, **Ecotoxicol. Environ. Saf.** 145, 42–49.

Wu M, Xu Y, Ding W, Li Y, Xu H (2016) Mycoremediation of manganese and phenanthrene by *Pleurotus eryngii* mycelium enhanced by tween 80 and saponin. **Appl Microbiol Biotechnol.**,100,7249–7261.

Xiangliang P, Jianlong W, Daoyong Z.(2005) Biosorption of Pb(II) by *Pleurotus ostreatus* immobilized in calcium alginate gel. **Process Bio Chem,** , 40,2799–2803

Xiangliang P, Jianlong W, Daoyong Z (2009) Biosorption of Co(II) by immobilised *Pleurotus ostreatus*. **Int J Environ Pollut** ,3,289–298.

Ying G.G., Kookana R.S.(2005) Sorption and degradation of estrogen-like-endocrine disrupting chemicals in soil. Environmental Toxicology and Chemistry,24(10),2640-2645

Zhu MJ, Du F, Zhang GQ, Wang HX, Ng TB. (2013)Purification a laccase exhibiting dye decolorizing ability from an edible mushroom *Russula virescens*. **Int Biodeterior Biodegrad.**4,33–39.

P.G. Department of Botany,
Dev Samaj College for Women, Ferozepur City,
Punjab
email : geetanjlikaushal555@gmail.com

7. Multidisciplinary Approch Towards Sustainable Development and Climate Change

Prof. (Dr.) Rakesh Daiya

Abstract

Multidisciplinary approaches are required to address the complex environmental problems of our time. Solutions to climate change problems are good examples of situations requiring complex syntheses of ideas from a vast set of disciplines including science, engineering, social science, and the humanities. Unfortunately, most ecologists have narrow training, and are not equipped to bring their environmental skills to the table with interdisciplinary teams to help solve multidisciplinary problems. To address this problem, new graduate training programs and workshops sponsored by various organizations are providing opportunities for scientists and others to learn to work together in multidisciplinary teams. Two examples of training in multidisciplinary thinking include those organized by the Santa Fe Institute and Dahlem Workshops. In addition, many interdisciplinary programs have had successes in providing insight into climate change problems including the International Panel on Climate Change, the Joint North American Carbon Program, the National Academy of Science Research Grand Challenges Initiatives, and the National Academy of Science. These programs and initiatives have had some notable success in outlining some of the problems and solutions to climate change. Scientists who can offer their specialized expertise to interdisciplinary teams will be more successful in helping to solve the complex problems related to climate change.

The environment is defined as the complex of physical, biotic, and chemical factors (such as living things, climate, and soil) that act upon an organism or an ecological community and ultimately determine its survival and form **(Merriam-Webster's dictionary).** The "environment" includes land, water, air, and the interrelationship which exists between these elements and human beings, other living creatures, microorganism, plants, and property **(Environmental Protection Act 1986)**

Sustainable development has become an important phenomena for all over the world to enhance the relationship between the environment and development and has penetrated into all aspects of our life, its a highly integrated concept.

There are now numerous sustainable development evaluation methods to evaluate sustainable development progress. Which one to use will depend on the resources, the goals and the stakeholders. Ideally the method selected and its indicators would themselves indicate how to achieve sustainable development. Such a choice can be difficult without a framework to help ensure the representation of the essential elements of sustainable development.

Concept of Sustainable Development : Importance

The concept of **sustainable** development has been there for decades and as a modern concept, it was originally brought forward by the Brundtland Report in 1987 in which it was simply defined as *"development that meets the needs of the present world without compromising the ability of future generations to meet their own needs"*. Sustainable development can be facilitated through five guiding principles, i.e. living within **environmental** limits, ensuring a strong, healthy and just society, achieving a sustainable economy, promoting good governance and utilising information communication technology as a social responsibility. This blog delves deeper into what the concept of sustainable development entails, why is it important as well as its major types and benefits.

The goals of sustainable development were first adopted by the **United Nations** Member States in 2015. The concept of sustainable development aims to encourage the use of products and services in a manner that reduces the impact on the environment and optimizes the resources in order to satisfy human needs. To understand why sustainable development is the need of the hour, take a look at the following key pointers that elucidate upon its importance:

- Development of non-polluting renewable energy systems
- Population stabilization
- Integrated land-use planning
- Healthy cropland and grassland
- Woodland and re-vegetation of marginal lands

- Conservation of biological diversity
- Control of pollution in water and of the air
- Recycling of waste and residues

Operationalisation of the concept of sustainable development requires the formulation of sustainability indicators and of techniques to assess scores on such indicators. A flexible approach is proposed, which incorporates several normative issues (including intergenerational welfare and substitution of manufactured capital for natural capital) and leads to a choice of location–specific sustainability constraints. Key parameters include classes of environmental resources, threshold levels, spatial level and time path. Assessing the score on this sustainability indicator involves the measurement of the difference between actual and normative levels of resource use. Depending on the measurement scale at which information becomes available, multi–criteria analysis may be useful to arrive at overall sustainability scores.

Multidisciplinary Approach

The term "multidisciplinarity" refers to the fact that it encompasses more than one branch of study or discipline. Studies in diverse domains are more effective when they are multi-sectoral and multi-dimensional. Environmental studies (EVS) include a wide range of information from various academic disciplines. This is what is indicated by the fact that environmental studies are multidisciplinary.

When the same topic is studied from the view point of more than one discipline is called Multidisciplinary approach. Sustainable development is a complex and multidimensional issue, and this concept has received multidisciplinary attention since its appearance in the Brundtland Commission on Environment and Dovelopment issued 'Our common future ' in 1987.(United Nations Brundtland report) . According to this report sustainable development is "Development that meets the needs of the present without compromising the ability of future generations to meet their own needs".

In 2015, 195 nations agreed with the United Nation that they can change the world for the betterby the year 2030.

- Eliminate Poverty

Pollution and climate change

- Erase Hunger
- Establish Good Health and Well-Being
- Provide Quality Education
- Enforce Gender Equality
- Improve Clean Water and Sanitation
- Grow Affordable and Clean Energy
- Create Decent Work and Economic Growth
- Increase Industry, Innovation, and Infrastructure
- Reduce Inequality
- Mobilize Sustainable Cities and Communities
- Influence Responsible Consumption and Production
- Organize Climate Action
- Develop Life Below Water
- Advance Life On Land
- Guarantee Peace, Justice, and Strong Institutions
- Build Partnerships for the Goals

A multidisciplinary approach can be analysed by economic ecological ethical social spatial and political aspects of sustainability. Requirement of multi disciplinary approach encompasses the complex environmental problems of our time from a vast set of disciplines including humanities, science, engineering and social sciences, culture ,technologies, natural resources and environmental fields. Sustainable Development is a comprehensive multidisciplinary process involving coordination and function.SD does not confute economic growth of poor countries, however, promotion of economic development should be further studied.

A multidisciplinary approach can be analysed by economic ecological ethical social spatial and political aspects of sustainability. Requirement of multi disciplinary approach encompasses the complex environmental problems of our time from a vast set of disciplines including humanities, science, engineering and social sciences, culture ,technologies, natural resources and environmental fields. Sustainable Development is a comprehensive multidisciplinary process involving coordination and function.SD does not confute economic growth of poor countries, however, promotion of economic development should be further studied.

Sustainable development must be coordinated with the carrying capacity of the environment, which should based on natural capital and ecosystem .Aim of it is imposing life quality and advancing society.Proper policy and legal system is necessary for implementation of sustainable development in which integrated decision making and public involvement should be included.

in the perspective of intergenerational and intra-generational equity sustainable development should not be considered for conflict and coordination of population ,resources, the environment and development in one country or one generation but also be considered for conflict and coordination coordination of population ,resources, the environment and development between different countries and generations.

The concept of sustainable development was first raised as "Development that meets the needs of the present without compromising the ability of future generations to meet their own needs". This definition was broadly accepted ,Consequently many researcher and International organizations studied sustainable development from different angles:

Viewpoint of Ecology and Sustainable Development

Sustainable development depends upon healthy ecosystem. Economic system and our quality of life depends upon continuous and adequqte functioning of the earth's ecological system. Existence and development of human beings are depend upon the water cycle, Bio geochemical cycle, biodiversity and well functioning ecosystem. Ecosystem can be defined as:

a. It provides get all resources for human activities .
b. It provides life support system- ecosystem also provide services such as purification of air and water, detoxification and decomposition of waste, moderation of earth's climate, moderation of floods and droughts
c. Provide humans marketable ecosystem goods such as seafood, forage, timber, many industrial products and soil and recovery of soil fertility.
d. Biodiversity including diversity of genes.
e. Generation and renewal of soil and soil fertility

Sustainable development must be established in stable functioning ecosystems, especially those subsystems that are vital to human existence. However, human activities have placed great pressure on the environment in recent years. Accordingly, those anthropogenic environmental problems, especially global environmental problems, are now seriously affecting the normal functioning of ecosystems and thereby affecting international economic and social development. These problems mainly include climate change, ozone depletion, ocean pollution, loss of biodiversity, desertification, soil degradation, destruction and degradation of freshwater and forests, and accumulation of persistent pollutants in the environment.

All of the above problems have destabilized the earth's biogeochemical system and thereby threatened sustainable development. To solve these problems, knowledge of a single subject is not sufficient. Multidisciplinary efforts are needed. Global climate change can be an example. Scientific evidence shows the impact on climate systems of anthropogenic activities and the cumulative effects, and thereby the impact on the biosphere and human beings. Economic analysis shows the probable costs and benefits of climate change and compares the cost effectiveness of different climate mitigation options. Development of technologies improves new energy-saving technologies and development of new energy sources, including new burning technologies such as Integrated Gasification Combined Cycle (IGCC) and some others that are still in the conceptual stages. Moreover, people have realized that these problems can not be solved merely by technological advances. From the viewpoint of social science, irrational production and consumption patterns under different societal conditions should be studied to provide new ideas on energy consumption and lifestyles. These new ideas are very important for solving the problems.

Similarly, study of and solutions to other environmental problems, especially global environmental problems such as ozone depletion, loss of biodiversity, and degradation of forests, also need multidisciplinary efforts.

Viewpoint of Economics and Sustainable Development

From the viewpoint of economics, sustainable development means a new way of thinking about the traditional concept of economic growth and capital, a new understanding about traditional economic measures of gross national product andstandard national accounting. Traditionally, attention to development has been constrained to activities within economic fields and has not taken into consideration the externality of economic activities—depletion and damage of economic activities to the environment and natural resources. One reason is that what economic growth was concerned with was only produced capital, while the importance of natural capital and human capital in economic development was overlooked. Actually, both natural capital and human capital are very important to economic development in addition to produced capital. For instance, natural capital provides raw materials and the capacity to eliminate wastes, while human capital proves to be a crucial part of capital and resources for development.

Past experiences shows that economic growth is necessary but not sufficient to development. If damage to ecosystems by economic activities is not taken into consideration, then the carrying capacity of biological systems will be destroyed and thereby sustainability of economic development will be damaged along with the damaged ecosystems. Consequently, one important step is to consider rational pricing of natural resources and calculation of marginal social costs of pollution on the basis of a new understanding of what constitutes capital. In traditional standard national accounting, the value of natural resources and ecosystems, cost of environmental pollution, and depletion of resources are not adequately considered due to its limitation to purely economic analysis. Currently, discussions focus increasingly on modification of accounting systems. This process does not concern only economics, but also involves multidisciplinary knowledge.

One example is resource pricing and economic accounting. Many economists, ecologists, and ethicists are involved in the discussion about resource pricing and economic accounting. They have broken down the traditional theory of economics and consider the issues of resource pricing, trying to make prices reflect the value of resources and damage to environment by resource exploitation. Of the theories of natural resource pricing, the theory of marginal cost pricing is well accepted. According to this theory, the resource price paid by users should equal the cost of natural resource exploitation and depletion that is now undertaken by society and the cost of corresponding environmental damage. To sum up, the full price should include cost for resource proving and exploitation, cost of environmental damage and recovery cost, etc. Employing this theory, a comparatively rational pricing system can be set up.

Currently, a series of research frontiers reflect the function of economics in sustainable development, including measurement and evaluation of natural resources, economics of sustainable agriculture, simulation of ecological-economic systems, policy implications of ecological-economic analysis, ecological-economic solutions to environmental degradation, ecological valuation, resources accounting, environmental rights, and environmental taxes. This research and its contribution to sustainable development should also be multidisciplinary.

Encompassing the elements of Physics, Chemistry, Medical Science, Agriculture, Geography, and Biology, Environmental Studies is a vast field of study. Not only restricted to **environmental conservation** and management of resources, but it also lays emphasis on understanding:

- Types of pollution and their harmful effects on living organisms
- Biodiversity, its types, and causes of degradation
- Deforestation and methods to increase forest cover
- Desertification
- Waste disposal and sewage treatment, etc

Three Pillars of Sustainable Development

The concept of sustainable development is rooted in three main pillars that aim to achieve inclusive growth as well as create shared prosperity for the current generation and to continue to meet the needs of future generations. These three pillars are **Economic**, Social and Environmental Development and are interconnected and reflect the goals of community development and social and environmental stability. Let's take a look at these the pillars of sustainable development in further detail:

Economic Sustainability

Economic sustainability strives to promote those activities through which long-term economic growth can be achieved without having a negative impact on the **environmental**, social, and cultural aspects

of the community. As a key facilitator for the concept of sustainable development, the basic fundamentals of economic sustainability are as follows:
- Finding effective solutions for hunger and poverty in the world in environmentally sound ways;
- Economics is the study of how societies use their resources (water, air, food, fuel, etc.) and when combined with the concept of sustainable development, it focuses on attaining economic growth which is only sustainable and simultaneously improves our quality of life and environment;

Climate Change and the Environment

Fundamentally, the increase in the consumption of fossil fuels has altered the composition of greenhouse gases in the Earth's atmosphere (IPCC, 2013). Earth's oceans, which absorb carbon dioxide, have warmed and become more acidic over time due to a greater ability to absorb gases (IPCC, 2013). This global oceanic warming has resulted in an average sea level rise of about ⅓ of a centimeter per year. This rate is expected to increase dramatically in the coming century (IPCC, 2013). In fact, experts estimate an average sea level rise of 28-98 centimeters by 2100, depending on future greenhouse gas emissions (IPCC, 2013). Severe precipitation and storms cause massive damage to forests, bodies of freshwater, and soils. Oceanic changes and precipitation are not the only climate change effects that impact the environment; extreme heat has contributed to widespread droughts on all landmass continents, as well as an increase in forest fires (IPCC, 2013). Plant species are unable to migrate or adapt quickly enough, via seed dispersal and other methods of propagation, to outpace these geographic changes. Nearly 70% of plant species in the Amazon Rainforest are at risk of extinction by 2100 (World Wildlife Foundation [WWF], 2018). Furthermore, the plant adaptation challenges at hand have major implications for food security across the globe. Climate change has harmful effects on the very ecosystems that all living things rely upon.

Multidisciplinary Approaches to Climate Change

Multidisciplinary approaches are required to address the complex environmental problems of our time. Solutions to climate change problems are good examples of situations requiring complex syntheses of ideas from a vast set of disciplines including science, engineering, social science, and the humanities. Unfortunately, most ecologists have narrow training, and are not equipped to bring their environmental skills to the table with interdisciplinary teams to help solve multidisciplinary problems. To address this problem, new graduate training programs and workshops sponsored by various organizations are providing opportunities for scientists and others to learn to work together in multidisciplinary teams. Two examples of training in multidisciplinary thinking include those organized by the Santa Fe Institute and Dahlem Workshops. In addition, many interdisciplinary programs have had successes in providing insight into climate change problems including the International Panel on Climate Change, the Joint North American Carbon Program, the National Academy of Science Research Grand Challenges Initiatives, and the National Academy of Science. These programs and initiatives have had some notable success in outlining some of the problems and solutions to climate change. Scientists who can offer their specialized expertise to interdisciplinary teams will be more successful in helping to solve the complex problems related to climate change.

Conclusion

Environmental problems are often called wicked problems because they are fraught with uncertainties and intertwined with many factors and multilevel scales (Head 2008; Head and Alford 2015; Tonkinwise 2015). These wicked problems cannot be treated independently by academic disciplines to inform problem-solving, decisionmaking, and policy-making. They carry a series of linked problems which cannot be resolved in isolation, where both the nature of the problem and the preferred solution are highly disputed (Head 2008; Head and Alford 2015). Head (2008) characterizes wicked problems as having high levels of complexity (of elements, subsystems, and interdependencies), high uncertainty (of risks, consequences of action, changing patterns), and high divergence and fragmentation (in viewpoints, values, strategic intention). Drawing from Björnberg and Hansson (2011) who used these five categories of judgments for climate change discourse, broadly speaking environmental problems can also be investigated by considering the following: • Evaluation (How the consequences of environmental problems should be evaluated?) • Timing (When should action be taken?) • Distribution (How should the benefits and burdens of the environment be distributed?) • Procedures (Who should be involved in environmental decision-making?) • Goal conflicts (How should goal conflicts in environmental management be dealt with?) While the environmental scientific discipline requires specialization, to address problems of societal importance such as environmental problems and sustainability, asking the questions above will require a multidisciplinary approach integrating the environment and society.

References

1. The Global Challenge for Government Transparency: The Sustainable Development Goals (SDG) 2030 Agenda
2. Teaching Sustainability: A Multidisciplinary Approach Yosef Jabareen
3. Multidisciplinary approaches to climate change questions By: Beth A. Middleton
4. Environment Aand Development- Vol. I - Multidisciplinary Approaches to New Pathways to Sustainable Development - Xueyang He and Kunmin Zhang
5. Climate Change and Health : An Interdisciplinary Exemplar Shanda L. Demorest and Teddie M. Potter
6. Multidisciplinary Nature of Environmental Studies
7. Extreme Events and Climate Change: A Multidisciplinary Approach
8. Federico Castillo (Editor), Michael Wehner (Editor), Dáithí A. Stone (Editor)

HOD Law
Maharishi Arvind University, Jaipur

Pollution and climate change

8. पर्यावरण संरक्षण के कानूनी पहलू

हरकेश मीणा

भारत में प्राचीन काल से ही पर्यावरण संरक्षण किया जाता रहा है। हमारे वेदों में भी पर्यावरण संरक्षण का उल्लेख किया गया है। हमारे ऋषि मुनियो ने प्राकृतिक शक्तियों को देवता मानकर पुजा की है। सुर्य, जल व नदियों को देव स्वरूप माना गया है। हमारी सस्कृति में केला, पीपल, तुलसी, बरगद, खेजडी आदि वृक्षो को देव तुल्य मानकर उनकी पूजा की जाती है। मध्य काल एवं मुगल काल में भी भारत में पर्यावरण के प्रति प्रेम बना रहा लेकिन ब्रिटिश शासन में अग्रेजो ने अपने स्वार्थ के लिए पर्यावरण का नाश करना शुरू कर दिया और वही से पारिस्थतिकीय असंतुलन की शुरूआत हुई। आजादी के बाद औधोगीकरण, जनसंख्या विस्फोट तथा पश्चिमी संस्कृति के अनुसरण के कारण पर्यावरण प्रदूषण की समस्या पैदा हुई। पर्यावरण को स्वच्छ रखने के लिए हमारी सरकार ने समय–समय पर अनेक प्रयास किये। हमारे संविधान के अनुच्छेद 48 ए में कहा गया है कि राज्य पर्यावरण की रक्षा करेगा साथ ही देश के वनो और वन्य जीवो की रक्षा करने का प्रयास करेगा। संविधान के अनुच्छेद 51 ए में कहा गया है कि प्राकृतिक पर्यावरण जिसमें वन, झील, नदी और वन्य जीव आते है उनकी रक्षा करेंगे तथा उनका संवर्धन करके प्राणी मात्र के प्रति दया भाव रखेंगे। इसके अलावा समय–समय पर अनेक एक्ट बनाये गये जैसे– फैक्ट्रीज एक्ट 1948, फोरेस्टस कन्जरवेशन एक्ट 1960, वाइड लाइफ प्रोटेक्शन एक्ट 1972, फोरेस्टस एक्ट 1980, पर्यावरण संरक्षण अधिनियम 1986, वाइड लाइफ प्रोटेक्शन एक्ट 1995, जैव विविधता संरक्षण अधिनियम 2002 में प्रावधान किया गया कि सरकार उन परियोजनाओ का पर्यावरणीय प्रभाव जांच करेगी जिनसे जैव विविधता को हानि पहुचने की आशंका हो, राष्ट्रीय जल नीति 2002 बनाई गई। राष्ट्रीय पर्यावरण नीति 2004 में कहा गया कि प्रत्येक मानव को स्वस्थ पर्यावरण का अधिकार है तथा इस नीति में स्थानीय संस्थाओ को पर्यावरण संरक्षण के लिए शक्तिशाली बनाने की बात कही गई, वन अधिकार अधिनियम 2006 में वन संरक्षण को बढावा देने की बात कही गई। भारत में विभिन्न एक्टो के माध्यम से पर्यावरण संरक्षण के लिए अनेक कदम उठाये गये।

पर्यावरण संरक्षण का तात्पर्य है अपने चारो ओर के वातावरण को संरक्षित करें तथा उसे जीवन के अनुकूल बनाये रखे। प्राचीन काल से ही भारत पर्यावरण के प्रति सजग रहा। हमारी संस्कृति में पर्यावरण के प्रति विशेष लगाव रहा है। हमारे वेदो में पर्यावरण संरक्षण के प्रति विशेष सजगता रखने की बात कही गई। हमारी संस्कृति में पेड पौधो की पूजा की जाती रही है। वेदों में पर्यावरण संरक्षण की बात कही गई। इनमें पर्यावरण संरक्षण को बढावा देने का जिक्र किया गया है। वेदो में जल, पृथ्वी, अग्नि, वनस्पति, अन्तरिक्ष तथा आकाश के प्रति श्रद्धा प्रकट

की गई है। जल जीवन का मूल तत्व है इसलिए वेदों में उनके महत्व पर प्रकाश डाला गया है। ऋग्वेद में कहा गया है कि जल में औषधीय गुण विद्यमान होते है इसलिए जल को अमृत के समान माना है। अतः इस स्वच्छ बनाये रखना चाहिए। निःसन्देह यह बात सही है। जल से पृथ्वी पर हरियाली छायी रहती है वातावारण सुहावना बना रहता है। समस्त प्राणियो का जीवन सुखमय व आनन्द पूर्ण बना रहता है। वायु के बारे में कहा गया है कि वायु जीवनदायनी शक्ति है। इसलिए इसका स्वच्छ रहना भी आवश्यक है। वेदों में वृक्ष पूजा का भी जिक्र है। वृक्षों को श्रद्धा से पूजने की बात कही गई है। पीपल, बरगद आदि वृक्षो का जिक्र किया गया है। ऋग्वेद में अग्नि को पिता के समान कल्याण करने वाला कहा गया है। ऋग्वेद में कहा गया है कि सम्पूर्ण परिवेश शुद्ध रहना चाहिए। नदी, पर्वत, वन, उपवन, गांव, नगर सब को विस्तृत एवं उत्तम परिसर प्राप्त होना चाहिए तभी जीवन का सम्पूर्ण विकास हो सकेगा। ऋग्वेद में सुर्य को पिता तथा पृथ्वी को माता के समक्ष माना गया है। अतः उनके प्रति आदर रखना चाहिए। वेदो में स्पष्ट निर्देश है कि प्रकृति के प्रति सदैव पूर्ण श्रद्धा रखे और आनन्दमय जीवन व्यतीत करने के लिए उससे पर्यावरण की अनुकूलता प्राप्त करते रहे। वेदों में प्राकृतिक तत्वों से अनावश्यक छेड़छाड़ करने के दुष्परिणामो की ओर इशारा किया गया है तथा कहा गया है कि पर्यावरण सन्तुलन को बिगड़ने के दुष्परिणाम सम्पूर्ण सृष्टि को उठाने पड़ेगे। हमारे ऋषि मुनियो ने भी पर्यावरण को संरक्षित रखने के उपदेश दिये है तथा पर्यावरण प्रदुषण से जो हानिकारक प्रभाव पड़ेगे उनकी ओर संकेत किया है। इसके अतिरिक्त हमारे पुराणो में भी कहा गया है कि वृक्षो की सेवा से सम्पूर्ण सृष्टि की सेवा का पुन्य कार्य सम्पन्न होता है। पुराणों में कहा गया है कि पेड़ो को जल देना चाहिए ताकि ये सुखे नही। इनके हरे रहने से इन पर आश्रित प्राणीयो को सुख मिलता है एवं पर्यावरण सुधरता है। गीता में भी भगवान कहते है कि वृक्षों में, मै पीपल अर्थात सन्देश दिया है कि वृक्षों को भगवान के समान मान कर पर्यावरण की रक्षा करनी चाहिए। हमारे ऋषि मुनियों ने भी कहा है कि यदि पृथ्वी की रक्षा करनी है तो वृक्षों और जल की रक्षा करनी होगी। जीवन में चार आश्रमो में से ब्रह्मचर्य आश्रम, वानप्रस्थ आश्रम एवं सन्यास आश्रम का सीधा सम्बन्ध पर्यावरण से है। विक्रमादित्य एवं अशोक के शासन काल में भी वन की रक्षा को सर्वोपरि माना जाता था। हिन्दु संस्कृति में घरों में तुलसी का पौधा लगाने को शुभ कार्य माना जाता है। पीपल को देवता समझ कर उसकी पूजा की जाती है। क्योंकि वह ऑक्सीजन देता है। वहीं वट पूर्णिमा व आंवला नवमी का पर्व मनाया जाता है। अथर्ववेद में बताया गया है कि आवास के समीप शुद्ध जल युक्त जलाशय होना चाहिए। हमारी संस्कृति में नदियों को मां के समान माना गया है। हमारी वैदिक संस्कृति तथा ऋषि मुनियो के कारण हमारा पर्यावरण शुद्ध रहा। मुगल काल में भी पर्यावरण के प्रति प्रेम बना रहा। मुगल शासको के द्वारा बड़े–बड़े बगीचे लगवाये जाते थे। वे पर्यावरण संरक्षण के प्रति सजग रहते थे लेकिन ब्रिटिश काल में अंग्रेजो ने अपने स्वार्थ के लिए पर्यावरण का नाश करना

Pollution and climate change

शुरू कर दिया तथा वही से पारिस्थितिकीय असन्तुलन की शुरूआत हुई। स्वतन्त्र भारत में पर्यावरण संरक्षण के लिए कई कानूनी उपाय किये गये। हमारे संविधान में पर्यावरण संरक्षण की बात कही गई।

भारतीय संविधान में मौलिक कर्तव्यों में यह अपेक्षा की गई है कि प्रत्येक नागरिक पर्यावरण को सुरक्षित रखने में योगदान देगा। संविधान के अनुच्छेद 21 में कहा गया है कि कानून द्वारा स्थापित बाध्यताओं को छोड़कर किसी भी व्यक्ति को जीवन जीने के अधिकार से वंचित नही रखा जायेगा। यह अनुच्छेद जीवन जीने का अधिकार देता है और इसमें पर्यावरण का अधिकार, बीमारियों व संक्रमण के खतरे से मुक्ति का भी अधिकार शामिल है। क्योंकि स्वस्थ वातावरण का अधिकार मनुष्य के जीवन जीने के अधिकार की प्रमुख विशेषता है। अनुच्छेद 21 के अनुसार स्वस्थ वातावरण में जीवन जीने के अधिकार को पहली बार 1988 में रूरल लिटिगेशन एण्ड एंटाइटलमेंट केन्द्र बनाम राज्य के मामले में सर्वोच्च न्यायालय ने पर्यावरण अधिनियम 1986 के तहत पर्यावरण सन्तुलन सम्बन्धी मामलो को ध्यान में रखते हुए अवैध खनन को रोकने के निर्देश दिये। 1987 में एम सी मेहता बनाम भारतीय संघ मामले में सर्वोच्च न्यायालय ने प्रदूषण रहित वातावरण में जीवन जीने के अधिकार को अनुच्छेद 21 के अर्न्तगत जीवन जीने के मौलिक अधिकार के रूप में माना है। 1993 में केरल उच्च न्यायालय ने ये फैसला दिया कि अभिव्यक्ति की स्वतन्त्रता किसी भी नागरिक को तेज आवाज में लाउडस्पीकर व अन्य शोर शराबा करने वाले उपकरण आदि बजाने की इजाजत नहीं देता है। संविधान के अनुच्छेद 19 (1) में प्रत्येक नागरिक को अपनी इच्छा अनुसार कार्य करने की स्वतंत्रता है लेकिन यह प्रतिबन्ध है कि कोई भी नागरिक ऐसा कार्य नही कर सकता जिससे लोगो के स्वास्थ्य पर प्रतिकुल प्रभाव पड़े। भारत के संविधान के अनुच्छेद 48 में कहा गया है कि राज्य पर्यावरण की रक्षा करेगा साथ ही देश के वनों और वन्य जीवो की रक्षा का प्रयास करेगा। संविधान के अनुच्छेद 51 ए में कहा गया है कि प्राकृतिक पर्यावरण जिसमें वन, झील, नदी एवं वन्य जीव आते है उनकी रक्षा करेगे तथा उनका संवर्धन करके प्राणी मात्र के प्रति दया भाव रखेगे।

पर्यावरण के संरक्षण के लिए समय–समय पर विभिन्न अधिनियम बनाये गये ताकि पर्यावरण सुरक्षित रह सके। जल (प्रदुषण निवारण एवं नियन्त्रण) अधिनियम 1974 से पूरे देश में लागु किया गया। इस अधिनियम के अनुसार एक केन्द्रीय प्रदूषण नियन्त्रण बोर्ड की स्थापना की गई। इस तर्ज पर राज्यो में भी प्रदूषण नियन्त्रण बोर्ड की स्थापना की गई। इस अधिनियम के अनुसार यदि कोई व्यक्ति जानबूझकर पानी में प्रदूषण फैलाने वाले पदार्थ डालता है तो वह अपराधी होगा तथा उसे कानून के अनुसार दण्ड दिया जायेगा। जल (प्रदुषण निवारण एवं नियन्त्रण) अधिनियम 1977 में केन्द्रीय व राज्य प्रदूषण बोर्डो को कई शक्तिया प्रदान की गई जैसे कि उन्हे किसी भी औधोगिक परिसर में प्रवेश करने का अधिकार होगा। जल में छोड़े जाने वाले कचरे के नमूने लेने का अधिकार होगा

तथा दोषी पाये जाने पर किसी भी औद्योगिक ईकाई को पानी व बिजली की आपूर्ति रोकने के साथ-साथ उनको बन्द करने के लिए भी बोर्ड कह सकता है।

वायु (प्रदूषण निवारण एवं नियन्त्रण) अधिनियम 1981— इस अधिनियम के पीछे जून 1972 में सयुंक्त राष्ट्र संघ द्वारा स्टाकहोम में आयोजित पर्यावरण सम्मेलन की भूमिका रही। इस अधिनियम की प्रस्तावना में कहा गया है कि पृथ्वी पर प्राकृतिक संसाधनों के संरक्षण के लिए उचित कदम उठाने होगे। इस अधिनियम में मुख्य रूप से मोटर गाड़ियों और अन्य कारखानो से निकलने वाले धुएँ और गन्दगी को नियन्त्रण करने सम्बन्धित प्रावधान है। इस अधिनियम के अनुसार केन्द्र व राज्य सरकार दोनो को निम्न शक्तिया प्रदान की गई:—

1. राज्य किसी क्षेत्र को वायु प्रदूषित क्षेत्र घोषित कर सकता है।
2. प्रदूषित क्षेत्रो में औद्योगिक क्रियाओ पर रोक लगा सकता है।
3. औद्योगिक इकाई स्थापित करने से पहले बोर्ड से अनापत्ति प्रमाण पत्र लेना अनिवार्य है।
4. जांच के लिए किसी भी इकाई में प्रवेश करने का अधिकार है।
5. प्रदूषित इकाईयो को बन्द करने का अधिकार होगा।

वन्य जीव संरक्षण अधिनियम 1972— कृषि, उघोग व शहरीकरण के कारण वनो का जो कटाव हुआ है। उससे जीव जन्तुओ की प्रजातियां लुप्त प्राय हो गई है। कुछ होने के कगार पर है। इन प्रजातियों को बचाने के लिए सरकार ने अनेक कदम उठाये है। 1952 में भारतीय वन्यजीव बोर्ड का गठन किया गया है। इसके अन्तर्गत अभ्यारणय बनाने का प्रावधान किया गया है तथा 1972 में वन्यजीवन संरक्षण अधिनियम पारित किया गया। इसमें लुप्त होती प्रजातियों को बचाने की व्यवस्था की गई। इसमें प्रावधान किये गये कि संकटग्रस्त वन्य प्राणियों की सूची बनाई जाये तथा उनके शिकार पर पाबंदी लगाई जाये। चिड़ियाधर व अभ्यारणयों में मूलभूत सुविधाएं अवश्य हो। लुप्त होती प्रजातियों को संरक्षित किया जाये। वन्यजीव परामर्श बोर्ड का गठन किया जाये तथा वन्यजीवन के लाभो की जानकारी शिक्षा के माध्यम से दी जाये। वन संरक्षण अधिनियम 1980— इस अधिनियम का मुख्य उद्देश्य वनो के विनाश को रोकना है। इसमें कहा गया कि वन सम्बन्धी योजनाँ इस प्रकार की हो कि वनो को संरक्षण मिले। जहा तक सम्भव हो सके वनो की कटाई रोकी जाये। पशुओ के लिए चारागाह का ध्यान रखा जाये।

ध्वनि प्रदूषण नियन्त्रण कानून— भारत में ध्वनि प्रदूषण को वायु प्रदूषण में शामिल किया गया है। पर्यावरण संरक्षण अधिनियम 1986 की धारा 6 में ध्वनि प्रदूषको सहित वायु तथा जल प्रदूषको की अधिकता को रोकने के लिए कानून बनाने का प्रावधान है। ध्वनि प्रदूषको को आपराधिक श्रेणी में मानते हुए भारतीय दण्ड संहिता की धारा 268 तथा 290 का प्रयोग करने का अधिकार पुलिस प्रशासन को दिया गया है। पुलिस प्रशासन को त्यौहारो और उत्सवों पर गलियों में संगीत नियन्त्रण करने का अधिकार है। पर्यावरण संरक्षण अधिनियम 1986 का मुख्य उद्देश्य है

Pollution and climate change

घातक रसायनो की अधिकता को नियन्त्रित करना व पारिस्थितिकी तन्त्र को प्रदूषण मुक्त करना। इस अधिनियम में 26 धाराऐं है जिन्हे चार अध्यायो में बाटा गया है। यह अधिनियम 19 नवम्बर 1986 से लागु किया गया। जिसके प्रमुख उद्देश्य निम्न प्रकार है:-

1. पर्यावरण का संरक्षण एव सुधार करना।
2. मानव, प्राणियो, जीवो व पादपो को संकट से बचाना।
3. पर्यावरण संरक्षण हेतु व्यापक विधि का निर्माण करना।
4. मानव पर्यावरण के स्टाकहोम सम्मेलन के नियमो को कार्यान्वित करना।

यह अधिनियम व्यापक कानून है इसके द्वारा केन्द्र सरकार के पास ऐसी शक्तिया आ गई है जिसके द्वारा वह पर्यावरण की गुणवत्ता के संरक्षण व सुधार हेतु उचित कदम उठा सकती है।

जैव विविधता संरक्षण अधिनियम 2002– भारत में पेड़ पौधो व जानवरो की हजारो प्रजातिया पाई जाती है। जैव विविधता को संरक्षित करने के लिए भारत में जैव विविधता संरक्षण अधिनियम 2002 पारित किया गया जोकि केन्द्र सरकार को निम्न दायित्व सौपता है।

1. उन परियोजनाओ का पर्यावरण प्रभाव जाँचना जिनसे जैव विविधता को हानि पहुचने की सम्भावना हो।
2. जैव तकनीकी से उत्पन्न प्रजातियों के जैव विविधता तथा मानव स्वास्थ्य पर पड़ने वाले नकारात्मक प्रभाव के लिए नियन्त्रण तथा उपाय सुनिश्चित करना।

राष्ट्रीय जल नीति 2002– राष्ट्रीय जल संसाधन परिषद ने अप्रैल 2002 में राष्ट्रीय जल नीति पारित की जिसकी महत्वपूर्ण बाते निम्न प्रकार है:-

1. इसमें कहा गया कि किसी भी जल परियोजना के निर्माण में शुरू से लेकर अन्त तक मानव जीवन पर उसके असर का मुल्यांकन करना।
2. जल संसाधनों के विकास और प्रबन्ध पर सरकार के साथ –साथ सामुदायिक भागीदारी सुनिश्चित करने की बात कही गई।
3. नदियों के जल संग्रहण क्षेत्र बनाने पर आम सहमति व्यक्त की गई।
4. जल बटवारे की प्रक्रिया में पेयजल को प्राथमिकता दी गई।
5. जल के उपयोग व बचत के लिए जनता में जागरूकता बढाने की बात कही गई।

राष्ट्रीय पर्यावरण नीति 2004– इस नीति की प्रस्तावना में कहा गया कि समस्याओ को देखते हुए एक व्यापक पर्यावरण नीति का आवश्यकता है। साथ ही वर्तमान पर्यावरणीय नियमो तथा कानूनो को वर्तमान की समस्याओ के सन्दर्भ में संशोधन करने की आवश्यकता है। राष्ट्रीय पर्यावरण नीति के मुख्य उद्देश्य निम्न प्रकार है:-

1. सतत विकास का केन्द्र बिन्दु मानव है।
2. स्थानीय संस्थाओ को पर्यावरण संरक्षण के लिए शक्तिशाली बनाना होगा।
3. प्रत्येक मानव को एक स्वस्थ पर्यावरण का अधिकार है।

4. विकास के अधिकार की प्राप्ति पर्यावरणीय जरूरतो को ध्यान में रखकर की जानी चाहिए।
5. संकटग्रस्त पर्यावरणीय संसाधनो का संरक्षण करना।
6. संसाधनो का न्यायोचित उपयोग सुनिश्चित करना।
7. आर्थिक तथा सामाजिक नीतियों के निर्माण में पर्यावरणीय सन्दर्भ को ध्यान में रखना।

वन अधिकार अधिनियम 2006—यह अधिनियम वन सम्बन्धी नियमो में एक महत्वपूर्ण दस्तावेज है जो कि 18 दिसम्बर 2006 को पास हुआ। यह अधिनियम जगलो में निवास करने वाले या वनो पर अपनी आजीविका के लिए निर्भर अनुसूचित जनजातियों के अधिकारों की रक्षा करता है। उन्हे पशु चराने तथा जल संसाधनों के प्रयोग का अधिकार देता है तथा उनको विस्थापन की स्थिति में पुनर्स्थापन का अधिकार देता है। यह अधिनियम स्थानीय लोगो को भूमि पर अधिकार प्रदान कर वन संरक्षण को बढावा देता है।

इस प्रकार हम प्राचीन काल से लेकर वर्तमान तक पर्यावरण संरक्षण को लेकर सजग रहे है लेकिन वर्तमान में पर्यावरण असन्तुलन को देखते हुए यह आवश्यक है कि हमें पर्यावरण को स्वच्छ रखने के लिए अधिक कानूनी उपाय करने होगे।

सन्दर्भ सूची :
1. डॉ. रामकुमार गुर्जर, डॉ. बी. सी जाट, पर्यावरण अध्ययन, पचशील प्रकाशन, जयपुर—2004
2. डॉ. दुर्गादास बसु, भारत का संविधान एक परिचय, 2015
3. डॉ. दयाशंकर सिंह यादव, पर्यावरण का समाजशास्त्र
4. डॉ. रामकुमार गुर्जर, डॉ. बी. सी जाट, मानव एवं पर्यावरण, पचशील प्रकाशन, जयपुर—2002
5. डॉ. सविन्द्र सिंह पर्यावरण भूगोल का स्वरूप, प्रवालिका पब्लिकेशन्स, इलाहाबाद—2020
6. राजस्थान पत्रिका, दैनिक भास्कर समाचार पत्र
7. https://hi.vikaspedia.in>rural-energy

सह आचार्य, राजनीति विज्ञान
श्रीमती नर्बदा देवी बिहानी राजकीय स्नातकोत्तर महाविद्यालय नोहर
हनुमानगढ़, राजस्थान
email : meenaharkesh705@gmail.com

9. Seed to Seedling Transmission and Phytopathological Effects of *Pseudomonas Aeruginosa* Causing Brown Soft Rot Disease in Onion

Laxmi Meena[1] and Laxmikant Sharma[2]

Abstract

Alwar red onions affected with brown soft rot disease caused by *Pseudomonas aeruginosa* were selected for the transmission and phytopathological studies. Seed to seedling transmission of bacterial pathogen was observed under pot and field experiments. The studied bacterial pathogen isolated from symptomatic onion seeds and seedling samples was identified using morphological, cultural and biochemical tests. Onion seeds were planted in pot and fields to study the phytopathological effects caused by pathogen. It was observed from experiments that bacterial pathogen has potential to reduce the seed germination as well as produced characteristic symptoms. Under pot experiment infected seeds showed 88% germination as compared to 100% of healthy seeds. Total loss was recorded 12% while 40% was recorded in field experiment. The seedlings with brown blight appearance die before maturity. Harvested onion bulbs showed typical brown spots and fleshy and sticky outer scales. The pathogen was recovered from harvested symptomatic onion plant parts showing seed-borne nature of the studied bacterial pathogen.

Keywords : Onion, bacteria, transmission, symptoms, disease.

Introduction

Onions are an important part of daily diet contains antioxidants and organic sulfur compounds. Besides as a good source of vitamin B6, vitamin C, potassium and manganese, they also provide little dietary fiber (Kochhar, 2016). The bulbs may help to reduce cholesterol level and help to fight inflammation protecting against heart diseases. It also strengthens immune system by providing sulfides and other phytochemicals. The production of onion has increased over the time with the availability of improved farming techniques and methods. However, quality of edible onion bulbs still falls short of the requirement (Meena et al. 2002). Like other vegetable crops, onion crop also suffer from many microbial pathogens and pests causing severe loss to the crop. Many fungal, bacterial pathogens and virus have been found associated with the crop are responsible to reduction of yield and quality. The major bacterial pathogens attacks to the crop are *Pseudomonas viridiflava, P. aeruginosa, P. syringae* pv. *syringae* and *Xanthomonas axonopodis* pv. *allii*. The present study investigated at *P. aeruginosa* causing brown soft rot disease in onion. The studied bacterial pathogen is soil-borne in nature responsible for significant crop yield and quality loss. The pathogen has wide hot range spread by the moisture and water splashes from diseased plants to healthy plants.

Materials and Methods

The study was undertaken in different blocks of Alwar district of Rajasthan, India. During field surveys and studies symptomatic and possible bacterial infected plant parts were collected and brought to the laboratory for further experiments. Upon incubation on nutrient agar (NA), these diseased plant parts oozes out bacterial colonies which were isolated and subjected to produce pure colonies on NA by restreaking. The isolated bacterial colonies were subjected to morphological, cultural and biochemical characterization. Morphological tests such as Gram's staining, KoH solubility test and catalase tests were performed. Bacterial colonies were also incubated on other media such as sucrose nutrient agar (SNA) medium and King's medium B agar (KmB) to observe the growth and colony characteristics. To identify bacterial colonies at species level, various biochemical tests i.e. Levan formation, oxidase test, potato oxidase test, arginine dihydrolase test, tobacco hypersensitivity tests, aesculin hydrolysis, cetrimide test were performed (Lelliot and stead, 1987, Schaad et al., 2001). To study the transmission and phytopathological effects of *P. aeruginosa* pathogen, infected and healthy onion seeds were sown in pots and fields for the development of symptoms. The data related to seed germination, seedling health, seedling length, symptomatic seedlings were recorded. Seedling vigor index was also calculated using formula seedling length* seed germination percentage.

Results and Discussion

During field surveys various symptoms of onion diseases were observed and symptomatic plant parts were analyzed for the presence of bacterial pathogen. Upon morphological, cultural and biochemical characterization, isolated bacterial pathogen was identified as *Pseudomonas aeruginosa* causing brown soft rot disease of onion. The bacterial pathogen was found soil borne as well as seed borne in nature as the seed to seedling transmission was observed in pot and field experiments (Table 1). The pathogen significantly reduced the seed germination and produced typical brown spot symptoms on raised seedlings. The seedling vigour index was also reduced to 480 as compared to 910 of healthy seedlings. In pot experiments 25 seed samples of each sample Paos-2112 (Symptomatic) and Paos-2108 (Healthy) were sown. Seed germination was observed 88% as compared to 100% for symptomatic seeds. Total loss was recorded 12% showing significant effects of bacterial infection. In the present study, results were found accordance with the study of Tiwari and Singh (2017) reported that maize seeds treated with *P. aeruginosa* results in inhibition and loss of seed germination and found it a major concern loss of crop yield. Chahtane et al. (2018) also reported loss of seed germination in Arabidopsis showing brown spot symptoms.

Table 1: Identification of bacterial isolates (*Pseudomonas aeruginosa*) isolated from onion seeds and plant samples.

S. No.	Characteristics		*Pseudomonas aeruginosa* (Isolate nos. Paos-2112)
1.	Gram stain reaction		Negative
2.	KOH test		Positive
3.	Catalase test		Positive
4.	Colonies on NA		Round, smooth, transleucent, fried egg appearance
5.	Colonies on KmB agar		Blue-green fluorescence under UV
3.	LOPAT	Levan formation on	Negative

		sucrose nutrient agar (SNA)	
		Oxidase test	Positive
		Potato soft rot test	Negative
		Arginine dihydrolase test	Positive
		Tobacco hypersensitivity response	Negative
5.	Gelatin liquefaction		Positive
6.	Indole production test		Negative
7.	Growth on cetrimide agar		Positive
8.	Citrate utilization test		Negative
9.	Pathogenicity test		Yellow lesions and water soaked spots

Reference
1. Kochhar, S. L., 2016. Economic botany fifth edition. Cambridge University Press. ISBN: 978131663822.pp 680.
2. Meena, M.L., Chauhan, M.S. and Rathore, S. 2002. Adoption and constraints in onion production technology in Alwar district of Rajasthan. Raj. J. Extn. Edu.10:88-92.
3. Lelliott, R. A. and Stead, D. E., 1987. Methods for the diagnosis of bacterial diseases of plants. In Methods in Plant Pathology, Vol. 2. Blackwell Scientific Publication, Oxford, London. pp 216.
4. Schaad, N. W., Jones, J. B. and Chun, W., 2001. Laboratory guide for identification of plant pathogenic bacteria. Third edition. Laboratory guide for identification of plant pathogenic bacteria. APS Press, 3340, Pilot Knob Road, St. Paul, MN, 55127-2097, USA. pp 373.
5. Tiwari, P. and Singh, J.S. (2017).A Plant growth promoting rhizosphere Pseudomonas aeruginosa strain inhibiting seed germination in Triticum aestivum (L.) and Zea mays (L.). Microbiology research. 8(7233):73-79.

6. Chahtane H, Nogueira Füller T, Allard PM, Marcourt L, Ferreira Queiroz E, Shanmugabalaji V, Falquet J, Wolfender JL, Lopez-Molina L. The plant pathogen *Pseudomonas aeruginosa* triggers a DELLA-dependent seed germination arrest in *Arabidopsis*. Elife. 2018 Aug 28;7:e37082.

**Dept. of Botany,
Raj Rishi Govt. College, Alwar (Rajasthan)**
email : [1]laxmi15meena@gmail.com,
[2]sharmalaxmikant999@gmail.com

10. Spectrophotometric Determination of Labetalol Using CDNB Reagent

B. Eswara Naik[1,4],
C. N. Rao[2]
C. Narasimha Rao[3*]

Abstract
A simple and sensitive spectrophotometric method has been developed for the determination of Labetalol (LBT) in pure and in pharmaceutical formulations. The developed method is based on the formation of pale yellow colored charge-transfer complex between the drug, LBT and 1-chloro-2,4-dinitrobenzene (CDNB) reagent. The formed complex showed maximum absorbance at 410 nm against blank. The limit of detection and quantitation were 0.1236 mg/ml and 0.4417 mg/ml respectively. The influence of commonly used exipients on the determination of LBT was studied. The linearity was observed between 6-28 µg/ml. The results of analysis were validated by recovery studies, accuracy, precision, LOD, LOQ, robustness and ruggedness, which indicated that the present method can be successfully applied for the determination of LBT in pure and in pharmaceutical formulations.

Keywords : Spectrophotometric method, labetalol, charge-transfer complex, CDNB.

1. Introduction
Labetalol (LBT) (Fig 1) is a non-cardiovascular β-blocker. It possesses some intrinsic sympathomimetic activity. It has α1 blocking properties which decrease peripheral vascular resistance. It is used to induce hypotension during surgery[1]. LBT is the subject of a monograph in each of the British Pharmacopoeia [2], the United States Pharmacopoeia[3] and the European Pharmacopoeia[4]. It is the first adrenergic antagonist capable of blocking both α and β receptors. It is a moderately potent hypotensive and is especially useful in pheochromocytoma. Beside these important pharmacological activities, Labetalol therapy exhibits hepatotoxicity and renal failure due to over dosage. Labetalol (LBT) is commonly recommended as a safe option for the treatment of maternal

hypertension during pregnancy[5]. In the literature, some spectrophotometric methods have been reported for the determination of LBT [6,7]. In the present work, we report a new sensitive, precise and accurate spectrophotometric method for the determination of LBT in pharmaceutical formulations and in biological fluid samples.

2. Experimental
2.1. Instrumentation
A Shimadzu UV-Visible spectrophotometer (UV-160A) with a matched pair of 10 mm quartz cell was utilized for all measurements. Mettler Toledo analytical balance (accuracy 0.1 mg) was used for weighing all the samples.

2.2. Materials and Reagents
LBT was procured from Sigma-Aldrich. Formulations were purchased from local market. All the used chemicals were analytical grade. Double distilled water is used throughout the experiment. A stock solution of LBT was prepared by dissolving accurately weighed 100 mg of pure drug in 100 ml standard flask by dissolving in methanol and sonicated to get required concentration of 1 mg/ml. From this, further dilutions were made with double distilled water to get required concentrations and used for current investigation.

2.3. Method Development
Standard drug solution was transferred into the clean and dried volumetric flasks. To each flask, 3 ml of 3% concentration of CDNB solution was added and the contents of all the standard flasks were heated at 98° ± 2°C. All these were cooled to room temperature and the formation of pale yellow colour was observed and the maximum absorbance was measured at 410 nm against the blank. The amount of drug was computed from calibration graph.

3. Results and Discussion
3.1. Absorption Spectrum
Different volumes of drug solution were taken into a clean and dry volumetric flask. For this solution, 3% of CDNB was added and entire contents were heated upto 98°±2°C and cooled at laboratory temperature. During this, change in the colour of solution was observed as pale yellow. Maximum absorbance was measured at 410 nm against blank reagent (Fig.2).

3.2. Effect of Reagent CDNB Concentration

Fresh aliquots of 3% CDNB solution was transferred into the fixed concentration of drug solution containing a series of volumetric flasks. Pale yellow colour was noticed after sometime on the addition of 3.0 ml of reagent and the same volume was considered for further investigation.

3.3. Effect of Concentration of the Drug

The effect of concentration on the absorbance of pale yellow coloured solution was studied and the linearity was observed in the range of 6-28 μg/ml by addition of fixed volume of reagent with a maximum absorbance at 410 nm against blank, calibration curve was constructed with obtained results and all results were obeyed the Beer's law.

3.4. Analytical Method Validation

The proposed analytical method was validated according to International conference on harmonization (ICH) guidelines[8,9] for the determination of the drug. The validation parameters such as linearity, accuracy, precision and specificity, Limit of detection (LOD), Limit of quantitation (LOQ), robustness and ruggedness were studied.

3.4.1. Linearity

The linearity of an analytical method is its ability to elicit test results that are directly or by a well defined mathematical proportional to the concentration of analyte in sample within a given range (Fig. 3). For studying the linearity of the method, different concentrations of the drug solution were prepared and calibration plots were constructed. From the plots drawn between the concentration of the drug and absorbance within the concentration range, the regression equations were computed. The optical characters like Beer's law limit, Sandell's sensitivity, molar absorptivity, are given in the Table 1.

3.4.2. Robustness and Ruggedness

To study the robustness of the proposed method, a few analytical parameters like concentration of the drug, concentration of the reagent and shaking time were interchanged. Even after that, it was noticed that the results were unaffected. The method of ruggedness is studied as the percentage of relative standard deviation for the present method developed by two different analysts from two different instruments in two different days and it was observed that there is no significant difference between the two analysts and instruments. Hence the developed analytical method is robust and rugged.

3.4.3. Accuracy

The accuracy of an analytical method is the close agreement between the accepted and obtained value. The obtained accuracy results proved that the recovery values in drug and in pharmaceutical formulations were within the acceptance criteria and details were presented in the Tables 2 and 3.

3.4.4. Precision

Precision of a method is a measure of the ability to create reproducible results. It is evaluated using six separate determinations for repeatability, precision and reproducibility. The intra and inter day precision were evaluated and found % RSD is less than 1.0 that proves that there is no considerable difference for the assay which is tested in inter-day and intra-day from pharmaceutical ingredients and results were presented in Tables 2 and 3.

3.4.5. Recovery

For recovery studies, the selected drug samples taken in various concentrations were analyzed by the proposed method and the recovery percentages were found to be more precise which shows the accuracy and selectivity of the proposed method. The average recovery results were summarised in the Table 4.

3.4.6. Specificity and Selectivity

To assess the developed of the method, the effect of excipients in its dosage forms like starch, lactose, glucose, sugar, talc etc. were studied. The results indicated that there was no interference from the excipients in its dosage forms. The results are shown in Table 5.

4. Applications

Blood and urine samples were collected from healthy donors, and centrifuged at 3000 rpm per min. for nearly 10 mins. The resulted solutions were filtered and preserved in the absence of light at a temperature of $4^{\circ}C$. From these solutions, various concentrations of the drug LBT were analyzed with the help of proposed analytical method and these results were recorded in Table 4. Hence, the proposed method can be successfully applied to recover LBT in biological samples, viz. urine and blood due to its high accuracy and good recoveries.

5. Conclusions

A simple, accurate and reliable spectrophotometric method for determination of LBT in pure and in pharmaceutical formulations was developed. By using this method it is possible to determine LBT with good precision and accuracy. The linearity of the calibration standards of the drug by the reported method was good from the result of correlation coefficient. The overall recovery of the drug by the proposed methods was satisfactory. LOD, LOQ, molar absorbity and Sandell's sensitivity values were calculated which indicated that the proposed analytical method was accurate, simple and reproducible for the estimation of LBT in pure and in pharmaceutical formulations.

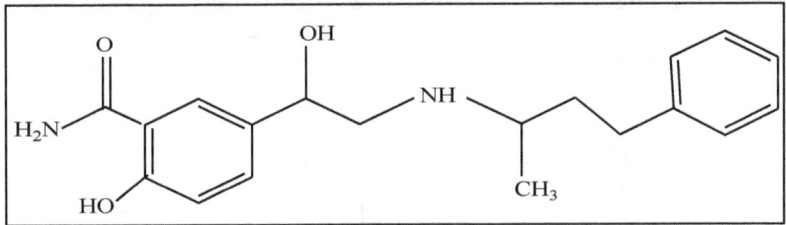

Fig. 1 Structure of Labetalol

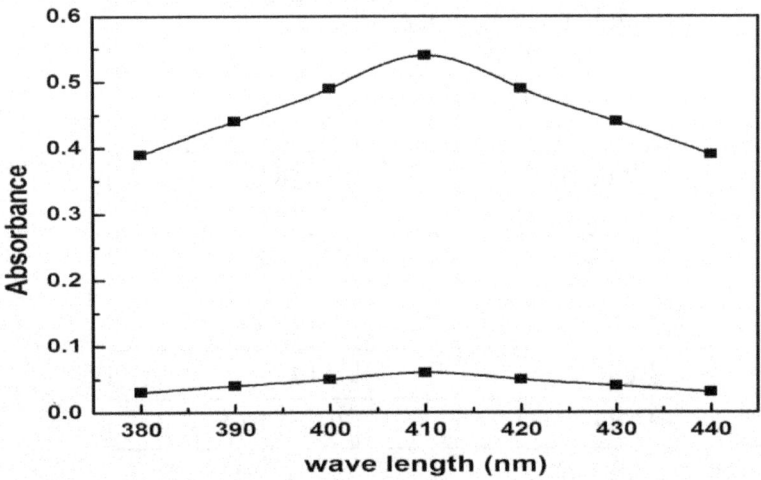

Fig. 2 Absorption spectrum of LBT with CDNB

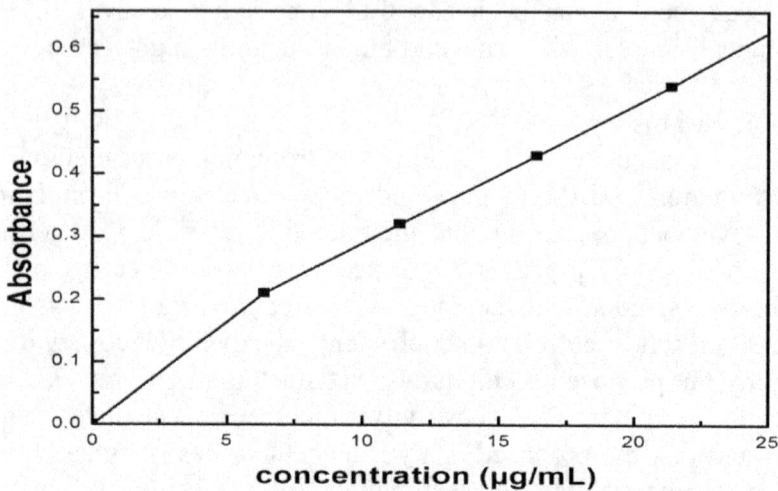

Fig. 3 Calibration plot of LBT

Table 1 Spectral characteristics of the drug with reagent

Parameter	
λ_{max} (nm)	410
Beer's law limit (μg/ml)	6- 28
Molar absorbance (L.mol^{-1} cm^{-1})	2.523×10^4
Sandell's sensitivity (μg.cm^{-2}/0.001 A.U)	0.0019
Correlation coefficient (r^2)	0.9967
Slope (m)	0.0284
Intercept (c)	0.2324
%RSD	0.1852
Colour	Yellow
LOD	0.1236
LOQ	0.4417

Table 2 Evaluation of accuracy and precision results of the proposed method in bulk form

Taken mg/ml	Intra day				Inter day			
	*Found mg/ml	Recovery %	± SD	% RSD	*Found mg/ml	Recover %	± SD	% RSD
2	1.98	99.17	0.006	0.29	1.98	99.00	0.017	0.87
4	3.96	99.00	0.010	0.25	3.96	99.08	0.006	0.15
6	5.96	99.28	0.006	0.10	5.97	99.50	0.026	0.44

Table 3 Evaluation of accuracy and precision results of the proposed method in pharmaceutical dosage form

Pharmaceutical formulation	Taken mg/ml	Intra day				Inter day			
		*Found mg/ml	Recovery %	± SD	%RSD	*Found mg/ml	Recovery %	± SD	% RSD
gravidol	4	3.96	99.08	0.012	0.29	3.93	98.17	0.015	0.39
labebet	6	5.97	99.44	0.012	0.19	5.95	99.17	0.010	0.17
trandate	8	7.96	99.50	0.017	0.22	7.94	99.21	0.035	0.44

*Average of six determinations

Table 4 Determination of recovery of LBT in pharmaceutical formulation

Name of the drug	Pharmaceutical formulation	Labeled amount (mg/ml)	*Found (mg/ml)	Recovery %	± SD	% RSD
Labetalol	gravidol	5	4.95	99.00	0.020	0.40
	labebet	5	4.97	99.40	0.010	0.20
	trandate	5	4.97	99.33	0.012	0.23

*Average of six determinations

Table 5 Determination of labetalol in presence of excipients

Excipients	Amount taken mg/ml	*Found mg/ml	Recovery %	±SD	RSD%
Glucose	5	4.96	99.13	0.012	0.23
Sucrose	10	9.96	99.57	0.006	0.06

Lactose	15	14.88	99.20	0.026	0.18
Dextrose	10	9.96	99.60	0.010	0.10
Talc	15	14.89	99.27	0.010	0.07
Starch	20	19.75	98.73	0.169	0.86

*Average of six determinations

Table 6 Method accuracy from recovery assay

Sample	Added mg/ml	*Found mg/ml	Recovery %	±SD	%RSD
Blood samples	0.2	0.197	98.67	0.001	0.29
	0.4	0.397	99.25	0.002	0.44
	0.6	0.596	99.39	0.002	0.35
	0.8	0.791	98.88	0.003	0.33
Urine samples	0.4	0.397	99.17	0.001	0.29
	0.6	0.597	99.44	0.001	0.10
	0.8	0.796	99.50	0.002	0.25
	1.0	0.987	98.67	0.006	0.59

References
1. F. Belal, S. Al-Shaboury, A.S. Al-Tamra, *Il Farmaco*, 2003, 58, 293.
2. The British Pharmacopoeia, HMSO, London (1993), vol. I and II, London, 1993, 978.
3. The United States Pharmacopoea XXIII, The USP Convention, Rockville, 1995, 866.
4. The European Pharmacopoeia, Council by Europe, Strasburg, 2001, pp. 1039 /1040.
5. Robert Rednic, Iasmina Marcovici, Razvan Dragoi, Iulia Pinzaru, Cristina Adriana Dehelean,
    Mirela Tomescu, Diana Aurora Arnautu, Marius Craina, Adrian Gluhovschi, Mihaela

Valcovici and Aniko Manea, In Vitro Toxicological Profile of Labetalol-Folic Acid/Folate Co-Administration in H9c2(2-1) and HepaRG Cells, Medicina, 2022, 58, 784.
6. Nafisur Rahman, Nishat Anwar, Mohammad Kashif, Md. Nasrul Hoda and Habibur Rahman, Determination of Labetalol Hydrochloride by Kinetic Spectrophotometry Using Potassium Permanganate as Oxidant, Journal of Mexican Chemical Society, 2011, 55, 105-112.
7. Nafisur Rahman, Habbiur Rahman, Syed Nazmul Hejaz Azmi, . Determination of Labetalol Hydrochloride in Drug Formulations by Spectrophotometry, Journal of Chinese Chemical Society, Taipei, 2007, 54, 185-196.
8. International conference of Harmonization (ICH) of Technical requirements for the registration of Pharmaceutical for human use, validation of Text and analytical procedures Q2(R1): Current Step 4 version, Parent Guideline dated 27 October 1994, (Complementary Guideline on Methodology dated 6 November 1996, Incorporated in November 2005).
9. ICH Q2A & Q2B. Analytical methods: A statistical perspective on the ICH Q2A and Q2B guidelines for validation of analytical methods. Support to Biopharm International. 2006.

[1]**Dept. of Chemistry,**
Rayalaseema University, Kurnool, Andhra Pradesh,
[2]**Department of Chemistry,**
S. V. University, Tirupati, Andhra Pradesh,
[3]**Dept. of Zoology,**
Govt. College for Men (A), Kadapa, Andhra Pradesh,
[4]**Dept. of Chemistry,**
IIIT-R.K.Valley, RGUKT-AP., Kadapa, Andhra Pradesh,
*Corresponding author email : narasimharao.svu@gmail.com

11. सतत आर्थिक विकास व पर्यावरण संरक्षण : अन्तर्निर्भरता तथा सामाजार्थिक लागतें

श्रीमती कविता शर्मा

1. शोध-सार (Abstract) :

सतत, वहनीय अथवा टिकाऊ विकास की अवधारणा पारिस्थितिकी तंत्र और आर्थिक विकास दोनों से समान रूप में जुड़ी है। चूंकि दोनों अन्तर्निर्भर हैं तो एक की गतिविधियों के पड़ने वाले प्रभावों से दूसरा अपने आपको बचा नहीं सकता। चूंकि मनुष्य के द्वारा किया गया आर्थिक विकास अब खुद उसी के जीवन के लिए खतरा बनता जा रहा है अतः विश्व स्तर पर सतत विकास की अवधारणा पर विचार - मंथन जारी है। यह रिसर्च आर्टिकल इस बात की चर्चा करता है कि आर्थिक विकास के नाम पर जो सहुलियतें हमने पैदा की, वही आज पर्यावरण और मानव स्वास्थ्य के लिए घातक बनकर चुनौतियां खड़ी कर रही है। यह इस बात पर भी ध्यान आकृष्ट करता है कि पूरा विश्व इसकी सामाजार्थिक लागतें उठा रहा है। इसके लिए विकसित तथा विकासशील अर्थव्यवस्थाएं क्या कर रही है ? क्या सतत विकास की अवधारणा को हकीकत के धरातल पर उतारा जाना संभव है ?

Keywords : आर्थिक विकास, पारिस्थितिकी-तंत्र, सामाजार्थिक लागतें, सतत विकास, अर्थव्यवस्थाएं

2. शोध संकल्पनाएँ :

(i) अर्थव्यवस्थाओं का आर्थिक विकास जरूरी है। इससे मनुष्य के कल्याण का स्तर ऊँचा होता है अर्थात् आर्थिक विकास और आर्थिक कल्याण में उच्च कोटि का सहसम्बन्ध पाया जाता है।

(ii) वैज्ञानिक प्रगति ने मशीनीकरण को अत्यधिक बढ़ा दिया है, जिससे जीवाश्म ईंधन समाप्ति के कगार पर है तथा प्रदूषणों का स्तर अधिकतम होने से 'ग्लोबल वार्मिंग' का खतरा तापमापी पर उच्च स्तर पर है।

(iii) सतत् आर्थिक विकास प्रक्रिया में शामिल अवधारणाओं को कोविड़-19 महामारी ने नकारात्मक रूप से प्रभावित किया है।

(iv) निर्वहनीय विकास को वहनीय, सतत् अथवा टिकाऊ न बनाया जा सका तो निकट भविष्य में पृथ्वी का वातावरण जीवन योग्य नहीं रह जाएगा।

3. शोध प्रविधियाँ :

1. गणितीय विश्लेषण (Formula based analysis)
2. द्वितीयक समंकों का प्रयोग - सारणी, चार्ट आदि।
3. स्थानीय, राष्ट्रीय व अन्तर्राष्ट्रीय स्तर के शोध

4. प्रस्तावना एवं परिचय :

"अज्ञानी होना गलत नहीं, पर अज्ञानी बने रहना गलत है।"

– स्वामी दयानन्द सरस्वती

Pollution and climate change

सतत अथवा टिकाऊ विकास, आर्थिक विकास को संरक्षित रखने की अवधारणा का नाम है। सतत विकास से हमारा अभिप्राय है कि जो विकास प्रकृति का दोहन अथवा अवशोषण करके हमने हासिल किया है, उसे कायम रख सकें, न केवल दीर्घकाल तक वरन् अनंत काल तक। स्वामी जी का उक्त वर्णित वाक्य अनायास ही इससे जुड़ गया है जिसका भावार्थ यह है कि जब तक हम अज्ञानी बने रहकर अंधाधुंध तरीके से प्रकृति का शोषण करके आर्थिक विकास करते गए, तब तक मान लें कि हमें ज्ञान नहीं था कि हम पर्यावरण व पारिस्थितिकी तंत्र को कितना नुकसान पहुँचा चुके हैं? लेकिन अब जबकि हमें पूरी दुनिया के बदलते पर्यावरण, ग्लोबल वार्मिंग, ग्रीन हाउस गैसों का बढ़ता स्तर आदि की जानकारी हो चुकी है, तब भी अज्ञानी बने रहकर प्रकृति व पर्यावरण को इतनी अधिक मात्रा में नुकसान पहुँचाते रहना बहुत गलत व निदंनीय है।

समस्या इतनी विकट है कि विश्व भर के पर्यावरणविद्, नीति निर्माता, शिक्षक तथा न्यायालय सभी इस बात का विचार मंथन कर रहे हैं ताकि विकसित एवं विकासशील दोनों प्रकार के देश अपनी आर्थिक विकास की प्रविधियों में सुधार लाए ताकि जल, भूमि, वायु व ध्वनि प्रदूषणों का स्तर, एक मानक स्तर पर रखकर हम धरती को पहुँचाए गए नुकसान की भरपाई भी करें तथा नए नुकसान भी न करें। वर्तमान में सभी देशों में निरंतर बढ़ता जा रहा तापमान, पिघलते ग्लेशियर, समुद्री खारे पानी की बढ़ती मात्रा, दूसरी तरफ पीने के पानी की दुर्लभता बढ़ती जा रही है। आज जितनी प्राकृतिक आपदाएँ घटित हो रही हैं, उससे अर्थशास्त्री माल्थस के सुप्रसिद्ध जनसंख्या सिद्धान्त की व्यावहार्यता सिद्ध हो रही है।

यद्यपि अन्तर्राष्ट्रीय और राष्ट्रीय स्तर पर बहुत संवाद– संगम कार्यक्रम चल पड़े हैं। विभिन्न देशों व जिम्मेदार अन्तर्राष्ट्रीय संस्थाओं के मंचों पर, फिर भी ये सिर्फ शुरूआती दौर में है। हमें प्रकृति को बचाने का संकल्प लेकर उस पर अडिग भी रहना होगा अन्यथा न प्रकृति बचेगी न हम रहेंगे।

5. शोध–साहित्य :

सतत–विकास की अवधारणा का प्रादुर्भाव नब्बे के दशक के आखिरी दौर में हो चुका था। तब इसे एस.डी. (S.D.) विचारधारा के रूप में जाना जाता था, जिसके तहत् अन्तर्राष्ट्रीय स्तर पर बहुत वार्ताएँ हुई और विकसित तथा विकासशील देशों ने एक–दूसरे पर आरोप–प्रत्यारोप लगाते हुए भी इस दिशा में सकारात्मक ढंग से काम करने का संकल्प लिया। वर्तमान में यह केवल एस.डी. अवधारणा नहीं रह गई है। जोकि 1992 में संयुक्त राष्ट्र संघ द्वारा संचालित UNFCCC (United Nations Framework Eouvention on Climate Change) कार्यक्रम से शुरू हुई वरन् आज इसे एड.डी.जी. (Sustainable Development Goals) अवधारणा 2030, के रूप में संयुक्त राष्ट्र संघ के एजेण्डे के रूप में अपना लिया गया है। ब्रटलैण्ड रिपोर्ट ने इन दोनों अवधारणाओं को जोड़ने पर बल दिया। एस.डी. (S.D.) तथा ए.डी.जी. (A.D.G.) दोनों अवधारणाओं को मद्देनजर रखकर विभिन्न अकादमियों तथा व्यक्तियों ने शोध–पत्र

प्रकाशित किए हैं ताकि पारिस्थितिकी-तंत्र को बचाने का बेहतर मॉडल तलाश किया जा सके।

6. आर्थिक विकास तथा पर्यावरण : अन्तर्सम्बन्ध, धारणीय विकास क्यों जरूरी है ?

हम जानते हैं कि आर्थिक विकास विभिन्न अर्थव्यवस्थाओं के संकल्पों से जुड़ा रहता है। एक प्रकार से यह उसी का प्रतिरूप माना जा सकता है। दूसरी तरफ पर्यावरण वह लिहाफ है– जो हमारे ऊपर आच्छादित है, जिसमें भूमि, जल, वायु तथा जीव-जन्तु से बना पारिस्थितिकी तंत्र जो जीवन के लिए जरूरी है। ये दोनों ही अन्तर्निर्भर है।

एक तरफ जहाँ पर्यावरण ही अर्थव्यवस्था के लिए महत्त्वपूर्ण संसाधन उपलब्ध कराता है। वहीं अर्थव्यवस्था की आर्थिक क्रियाएँ भी पर्यावरण को प्रभावित करती है। इसी अन्तर्निर्भरता के रिश्ते में जब असंतुलन आ गया, उसके दुष्परिणाम प्रदूषण के विभिन्न प्रकारों के बढ़ते स्तरों– ओजोन परत में छेद, ग्रीन हाउस गैसेज का बढ़ता स्तर और इन सबके सामने हारती जिन्दगी हम देख पा रहे हैं। मतलब यह कि जो विकास हमने पारिस्थितिकी तंत्र को नुकसान पहुँचाकर किया है, वह धारणीय अथवा सतत नहीं है। विकास को नियमित व नियंत्रित करके इसकी धारणीयता (Sustainbility) को बढ़ाना होगा।

इसे हम संयुक्त राष्ट्र संघ (U.N.O.) के द्वारा जारी आँकड़ों से इस प्रकार समझ सकते हैं कि पिछले 50 वर्षों में जहाँ जनसंख्या में 2.4 गुना वृद्धि हुई वहीं उपभोग में 6 गुना से ज्यादा बढ़ोतरी भी हुई है, परन्तु चिन्ता का विषय यह है कि उपभोग में यह बढ़ोतरी सभी लोगों में समान रूप से वितरित नहीं हुई। U.N.O. के उपलब्ध आँकड़ों के अनुसार विश्व के सर्वाधिक 20 प्रतिशत अमीर लोग कुल उपभोग का 86 प्रतिशत हजम कर जाते हैं और शेष 14 प्रतिशत दुनिया की 80 प्रतिशत आबादी के हिस्से आता है। इससे असमानता के साथ-साथ गरीबी की खाई निरंतर गहरी होती जा रही है।

आर्थिक विकास का जो मॉडल हमने अपना रखा है उससे असमानताएँ बढ़ी हैं, जिससे अधिकतर आबादी की जीवन की गुणवत्ता पर प्रतिकूल असर पड़ा है। चूंकि धारणीय विकास जिसे हम सतत विकास कहते हैंएक बहुआयामी विचार है जिसके प्रभाव केवल आर्थिक विषमता के पैमाने से ही नहीं तोले जाते वरन् इसके सामाजिक व पर्यावरणीय, फलस्वरूप राजनीतिक आयाम भी घटित होते हैं। इसे सबसे अधिक सरल शब्दों में समझा जा सकता है पर्यावरण एवं विकास पर विश्व आयोग प्रतिवेदन 1987 में गई परिभाषा द्वारा कि, ''सतत विकास वह विकास होगा जो वर्तमान आवश्यकताओं को, भावी पीढ़ी की आवश्यकताओं को संतुष्ट करने की क्षमता से, समझौता किए बगैर संतुष्ट करता है।''

इसकी व्याख्या हेतु मेरे दिमाग में वो विज्ञापन आता है जिसमें एक मैकेनिकल इंजीनियर लाल बत्ती पर रूकी कार में बैठा अपने 12 साल के बेटे से पूछता है कि ''बेटे! बड़े होकर कैसी कार डिजायन करोगे ? तो वह जबाव देता

है, "मैं तो साईकिल डिजाइन करूँगा न पापा, क्योंकि आपकी पीढ़ी पैट्रोल तो छोड़ेगी ही नहीं हमारे लिए।"

बस यही करना है हमें, विकास को धारणीय या सतत बनाकर ताकि आने वाली पीढ़ियों के लिए संसाधन बचे रहें, जीवन बचा रहे।

7. सतत विकास के घटक :

यह प्रसन्नता का विषय समझा जा सकता है कि अब पूरी दुनिया सतत विकास के मुद्दे पर बात करना चाहती है और सचेत हो रही है। इसका प्रमाण है 2015 में U.N.O. की सामान्य महासभा की बैठक में विश्व स्तर पर 2030 तक सतत विकास की अवधारणा को प्राप्त करने का संकल्प लिया गया। जिसमें मुख्य 17 लक्ष्य निर्धारित किए गए जो कि सतत विकास के घटक या पैमाने कहे जा सकते हैं। इन्हें short में SDGs कहते हैं। ये निम्नलिखित प्रकार से हैं :—

Goal 1	:	No Poverty
Goal 2	:	Zero Hunger
Goal 3	:	Good Health and well being
Goal 4	:	Quality Education
Goal 5	:	Gender Equlity
Goal 6	:	Clean Water and Sanitation
Goal 7	:	Affordable and Clean Energy
Goal 8	:	Decent Work and Economic Growth
Goal 9	:	Industry, Innovation and Infrastructure
Goal 10	:	Reduced Inequality
Goal 11	:	Sustainable Cities and Communities
Goal 12	:	Responsible Consumption and Production
Goal 13	:	Climate Action
Goal 14	:	Life Below Water
Goal 15	:	Life on Land
Goal 16	:	Peace and Justice strong Institutions
Goal 17	:	Partnerships to Achieve the Goal

सन् 2016 प्रथम वर्ष था जब महासभा की बैठक में सम्मिलित 163 देशों ने 2030 तक इन लक्ष्यों की प्राप्ति की दिशा में अपने यहाँ नीतिगत प्रयास शुरू किए। bUgksaus # Envision- 2030 के नाम से इस एजेण्डे को गति प्रदान की, ताकि 2030 तक इन संकल्पों को पूरा कर दुनिया को जीने के लिए बेहतर जगह बनाया जा सके।

8. सतत विकास मुद्दे पर भारत : दृष्टिकोण व यथास्थिति :

भारत दुनिया में अब कोई अपरिचित देश नहीं रहा है। सबसे बड़े लोकतंत्र और दुनिया की आबादी के हिसाब से हम दूसरे स्थान पर काबिज हैं। जाहिर है

Pollution and climate change

यदि हमारे पास भरे जा सकने वाले पेट अधिक हैं तो उनके लिए काम करने वाले हाथ भी अधिक हैं। अपने उत्पादन को खपाने के लिए दुनिया के बाकी देश हमारी तरफ देखते हैं। हमने उत्पादन व उपभोग के लिए हमारी अर्थव्यवस्था के दरवाजे उनके लिए खुले मन से खोल रखे हैं। भारत पिछले 50 वर्षों से विकसित देशों की कतार में खड़ा होने का स्वप्न देख रहा है। परन्तु दुनिया के हर मंच से भारत ने सतत विकास की अवधारणा पर काम करने के लिए दुनिया के देशों को आवाज दी है। हमारे वर्तमान प्रधानमंत्री श्री नरेन्द्र मोदी ऐसी सम्मिटों में शामिल होते रहे हैं और SGDS 2030 के प्रति एक जागरूक रवैया रखते हैं। उनके इस कथन में इसकी झलक दिखाई देती है,

"एजेंडा 2030 के पीछे की हमारी सोच जितनी ऊँची है, हमारे लक्ष्य भी उतने ही समग्र हैं। इनमें उन समस्याओं को प्राथमिकता दी गई है जो पिछले कई दशकों से अनसुलझी हैं और इन लक्ष्यों से हमारे जीवन को निर्धारित करने वाले सामाजिक, आर्थिक और पर्यावरणीय पहलुओं के बारे में हमारी विकसित होती समझ की झलक मिलती है। मानवता के 1/6 हिस्से के सतत् विकास का विश्व और हमारी सुन्दर पृथ्वी के लिए बहुत गहरा असर होगा।"

जब मोदी जी अपने भाषण में उक्त SGDS 2030 लक्ष्यों के प्रति प्रतिबद्धता जाहिर कर रहे थे तो संयुक्त राष्ट्र महासभा के सामान्य सचिव श्रीमान् एंटिनियो गुटरेस ने भी सतत् विकास के इन लक्ष्यों के प्रति जागरूकता बढ़ाने की बात कही, उनके कथनानुसार,

"2015 में अनुमोदित 2030 SGDS एजेण्डा और उसके 17 सतत् विकास लक्ष्य इन चुनौतियों और इनके अंतर संबंधों के समाधान के लिए सम्पूर्ण और सामंजस्यपूर्ण फ्रेमवर्क प्रदान करते हैं। इनके अन्तर्गत सदस्य राष्ट्रों को सतत् विकास के सामाजिक, आर्थिक और पर्यावरणीय पहलुओं का समाधान संतुलित ढ़ंग से करना होगा। इन पर अमल करते हुए समावेशन और एकीकरण तथा किसी को पीछे छूटने न देने के सिद्धान्तों का पालन अनिवार्य होगा।"

सतत् विकास के 17 लक्ष्य SGDS 2030 को प्राप्त करने के लिए हमारी सरकारें कृत संकल्प हैं, लेकिन उसके अनुरूप जितने सकारात्मक परिणाम दिखाई देने चाहिए थे, उतने परिलक्षित होते नहीं दिखते।

Global Sustainable Development Report 2022 के अनुसार दुनिया के 163 देशों के 2030 तक प्राप्त कर सकने वाले 17 लक्ष्यों की दौड़ में भारत का स्थान पिछली रैंकिंग से गिरकर 121 हो गया। 2020 में भारत 117वीं रैंक पर था, 2021 में 120वीं रैंक पर गिरा और लगातार 2022 में यह रैंकिंग गिरकर 121 पर आ चुकी है।

भारत सरकार की सर्वेक्षण रिपोर्ट और Global Sustainable Development Goals 2030 के लिए प्रयासरत U.N.O. की रिपोर्ट के अनुसार 2019–2020 में इसका मुख्य मुख्य कारण भारत का इनमें से SGD-13 लक्ष्य पर पिछड़ना रहा, जो कि मुख्यतः Climate Change के प्रति सरकार द्वारा उठाए

Pollution and climate change

गए कदम थे। भारत के आठ राज्यों में इसके प्रति बेरूखी पाई गई। ये थे – बिहार, तेलगांना, राजस्थान, उत्तर प्रदेश, कर्नाटक, आन्ध्र प्रदेश, पंजाब तथा झारखण्ड। इन्हीं राज्यों के व्यक्तिगत लक्ष्य, SGD 2030 के लक्ष्य नम्बर 13 के साथ तालमेल सकारात्मक रूप से नहीं बैठा पाए, जिस कारण कुल 17 में से 11 लक्ष्यों पर बहुत बेहतर उपलब्धि हासिल करके भी शेष 5 या 6 लक्ष्यों में नकारात्मक अथवा गिरती हुई रैंकिंग के कारण भारत की SDG 2030 की रिपोर्ट में हम पिछड़ गए। निम्न तालिका में प्रथम 10 देशों और भारत तथा उसके पड़ौसी देशों की रैंकिंग दर्शाई गई है :

Rank	Country	Score (out of 100)
1	Finland	86.51
2	Denmark	85.63
3	Sweden	85.63
4	Norway	82.35
5	Austria	82.32
6	Germany	82.18
7	France	81.24
8	Switzerland	80.79
9	Ireland	80.66
10	Estonia	80.62
56	China	72.38
60	Singapore	71.72
70	Bhutan	70.49
76	Sri Lanka	70.03
82	Indonesia	69.16
98	Nepal	66.18
104	Bangladesh	64.22
121	India	60.32
125	Pakistan	59.34

Source : Global sustainable development Report - 2022

तालिका को देखें तो स्पष्ट है कि भारत अपने पड़ौसियों में भी Ranking में नीचे हैं, जबकि भारत एक बड़ा देश है तो उसके पास संसाधन व मशीनरी भी अधिक है। सिर्फ पाकिस्तान ही अकेला पड़ौसी देश है, जिसकी हमसे भी कम रैंकिंग 125वीं है।

9. अंधाधुंध विकास की सामाजार्थिक लागतें :

आर्थिक विकास की केवल आर्थिक लागतें ही नहीं होती जो कोई देश अपने संसाधनों का दोहन करके उठाता है। वरन् उसकी बहुत सी सामाजिक व

Pollution and climate change

मानवीय, लागतें भी होती हैं। पर्यावरणीय लागतों को स्वास्थ्य सुविधाओं की उपलब्धता तथा मरीजों की संख्या का अस्पतालों तथा स्वास्थ्य कर्मियों की संख्या से अनुपात आदि को यदि आर्थिक लागतों में ही सम्मिलित कर दें तो शेष बची लागतों को सामाजिक लागत कहा जा सकता है।

इसे इस प्रकार समझ सकते हैं, जैसे एक व्यक्ति जब सुबह अपने घर से कार्य स्थल के लिए निकलता है तो रास्ते में टूटी-फूटी सड़कें, बदबूदार कचरे के ढ़ेर, धुँआ छोड़ती गाड़ियाँ, दूर-दूर तक हरियाली का अभाव आदि सब उसके कार्य स्थल पहुँचने तक उसकी मनःस्थिति को नकारात्मक रूप से प्रभावित करते हैं। जिससे उसकी कार्यक्षमता पर प्रतिकूल असर पड़ता है। कार्य स्थल पर स्वास्थ्यजनक माहौल न होने के कारण उसकी दीर्घकालीन उत्पादकता पर असर पड़ता है। इस तरह सामाजिक व आर्थिक दोनों लागतें एक देश के विकास को सतत बना नहीं रहने देती। आज न केवल विकासशील वरन् विकसित देश भी इससे उत्पन्न होने वाली समस्याओं से जूझ रहे हैं, क्योंकि वे सिर्फ भूमि को सीमाओं में बाँध सकते हैं। हवा एवं कुछ हद तक जल को नहीं।

आधुनिक अर्थशास्त्रियों ने इसकी भयवहता को समझते हुए सामाजिक लागत-लाभ विश्लेषण के आर्थिक मॉडल विकसित किए हैं। जिसमें आर्थिक विकास की वास्तविक लागत निकाली जा सके और उसे मौद्रिक लाभ में से घटाकर वास्तविक लाभ ज्ञात किया जा सके। इन अवधारणाओं के पीछे वे अर्थशास्त्री हैं जो पर्यावरणविद भी रह चुके हैं, का खास योगदान रहा है, जिनमें जेम्स टॉबिन और सेम्यूलसन आदि के नाम लिए जा सकते हैं, जिन्होने दो अवधारणाऐं विकसित की हैं :–

(i) NEW → Net Economic Welfare
(ii) MEW → Measurement of Economic Welfare

आर्थिक विकास दुनिया के देशों की जरूरत है, जिनमें विकासशील देशों को विशेषकर, क्योंकि विकसित देश बहुत अधिक क्षमता विकसित कर चुके, प्रकृति को बहुत अधिक नुकसान भी पहुँचा चुके और अब वे इसे रोकना चाहते हैं। लेकिन विकासशील देशों की अभी ये जरूरत हैं। अतः वे प्रतिबन्धों के साथ रूकना नहीं चाहते तो अर्थशास्त्रियों द्वारा परियोजनाओं के लागत लाभ विश्लेषण मॉडलों को अपनाकर पारिस्थितिकी तंत्र का नुकसान कम करना संभव हो सकता है।

एक सामान्य उदहरण से अर्थशास्त्र का अल्प-ज्ञान रखने वाला पाठक भी इसे समझ सकता है :–

माना दो परियोजनाएँ हैं। किसी निवेशक या उद्यमी के समक्ष जो हैं 'X' तथा 'Y' जिनकी लागतों को C और लाभों को B कह सकते हैं।

(C = Cost & B = Benefits)

निवेशक का उद्देश्य है उसकी लागत न्यूनतम हो और लाभ अधिकतम हो, लेकिन यदि इसमें सतत विकास लक्ष्यों को शामिल किया जाए, पर्यावरणीय चिंताओं को डाला जाए तो इन परियोजनाओं का शुद्ध सामाजिक लाभ (Net-

Pollution and climate change

Social Benefit) ज्ञात करके उसे अधिकतम करना होगा। चूंकि विकास की परियोजनाएँ दीर्घकालीन होती है और दीर्घकाल तक लाभ देने वाली होती है। इसलिए आज की वर्तमान अवधि में भावी वर्षों के लाभों तथा लागतों की कटौती करके वर्तमान मूल्य (Present Value) का पता किया जा सकता है। इसे उद्यमी निम्न सूत्र से ज्ञात कर सकते हैं :—

$$NSB = PV(B-C) = \sum_{t=0}^{H} \frac{B_t}{(1+r)^t} - \sum_{t=0}^{H} \frac{C_t}{(1+r)^t}$$

यहाँ NSB = Net Social Benefit है।
B = Benefits
C = Cost
r = Discount rate
t = time (present)
H = Duration

U.S. Sub-committee on benefits and costs ने इस विश्लेषण पर चार कसौटियों का विवेचन किया है :—

(a) $B - C$
(b) $B - \frac{C}{I}$
(c) $\frac{\Delta B}{\Delta C}$
(d) $\frac{B}{C}$

यहाँ B = Benefits
C = Cost
I = Direct Investent
ΔB = Change in benefits between two periods
ΔC = Change in costs between two periods

उक्त चारों में से सर्वाधिक व्यावहारिक चौथी कसौटी $\frac{B}{C}$ है, जिसे लाभ–लागत अनुपात कहा जाता है। ऐसे में परियोजना अपनायी जाएगी अथवा नहीं अपनायी जाएगी– ये निम्न मूल्यों पर निर्भर करता है।

(i) यदि $\frac{B}{C} = 1$ है तो परियोजना सीमांत मानी जाएगी तथा केवल अपनी लागत को पूरा करेगी।

(ii) यदि $\frac{B}{C} > 1$ है तो लाभ, लागतों से अधिक होंगे और परियोजना अपनायी जाएगी।

(iii) यदि $\frac{B}{C} < 1$ है तो लाभ, लागतों से कम होंगे और परियोजना अस्वीकृत की जायेगी।

Pollution and climate change

इस प्रकार हम देखते हैं कि आर्थिक विकास की चुकाई जाने वाली सामाजिक लागतों को कम किया जा सकता है। यदि उसका वास्तविक मूल्य निकाला जाए। यह बिल्कुल सरल है समझना कि सतत विकास बनाए रखना है तो केवल वणिक बुद्धि से काम नहीं चलेगा वरन् साथ में पर्यावरणविद् बनकर सोचना पड़ेगा।

10. जनसंख्या वृद्धि का पर्यावरण संरक्षण पर प्रभाव :

पर्यावरण को समूची विश्व अर्थव्यवस्थाओं ने जितना नुकसान पहुँचाया है, उसका कारण आर्थिक विकास को ठहराया जाता है। लेकिन आर्थिक विकास किसके लिए हुआ है ? इसका उत्तर है बढ़ती हुई जनसंख्या जो पिछली सदी में सर्वाधिक तेज दर से बढ़ी। सन् 1920 में विश्व जनसंख्या का आँकड़ा दो अरब था, जो कि 2020 में 8 अरब को पार कर गया है तथा सन् 2050 में यह आँकड़ा दस अरब हो जाने का अनुमान है। अर्थशास्त्री माल्थस के जनसंख्या के सिद्धान्त के अनुसार यह संख्या विनाश की और जाति मानव जाति का परिचय कराती है।

इतनी बड़ी विश्व जनसंख्या के लिए संसाधन जुटाने हेतु हमने भारी मात्रा में पृथ्वी के पारिस्थितिकी तंत्र का दोहन किया और उसके घातक परिणाम हमारे समक्ष खड़े हैं जिनका जिक्र बार–बार अपने इस शोध–पत्र में मैंने किया है। संक्षेप में जनसंख्या के बढ़ते घनत्व, जिसे सामान्य भाषा में हम धरती पर बढ़ता बोझ कहते हैं, के कारण निम्न प्रभाव पड़ते हैं :–

(1) अत्यधिक अपशिष्ट पदार्थों का उत्सर्जन
(2) जैव–विविधता के लिए खतरा बन गया है। कई प्रजातियाँ विलुप्त होकर इतिहास के पन्नों में चली गई हैं।
(3) बढ़ते औद्योगिकरण के प्रभाव– जिससे अत्यधिक मात्रा में खतरनाक गैसों का बहाव वायु में तथा विषैले पदार्थों का उत्सर्जन नदियों में किया जा रहा है।
(4) सुरसा के मुँह की तरह बढ़ता शहरीकरण जो समस्त समस्याओं की जड़ है। औद्योगिकरण के संकेन्द्रण बने ये शहर अधिक वायु, जल व ध्वनि प्रदूषण करते हैं तो यहाँ रहने वाले लोगों का स्वास्थ्य भी प्रभावित होता है।
(5) वनों का घटता प्रतिशत भी जनसंख्या वृद्धि के पर्यावरण पर पड़ने वाले नकारात्मक प्रभावों को दर्शाता है।
(6) जलवायु में ग्रीन हाऊस गैसों का निरंतर बढ़ता स्तर भी इसी कारण है।
(7) भूमि का निरंतर क्षरण भी बढ़ती जनसंख्या के कारण बढ़ता है। ग्लोबल वार्मिंग के बढ़ते स्तर की वजह से धरती का तापमाप निरंतर बढ़ रहा है। ग्लेशियर पिघल रहे हैं, खारे पानी का जल स्तर बढ़ रहा है। प्राकृतिक आपदाएँ जैसे अकाल, अतिवृष्टि, अनावृष्टि आदि धरती की मृदा की ऊर्वरा का हनन कर रहे हैं। वर्तमान में ही इटली में पड़ रहे सूखे पर वहाँ की सरकार चिंतित है। ब्रिटेन और यूरोप में तापमान के 40 डिग्री तक पहुँच जाने से वहाँ की सरकारें चिंतित हैं। ये सभी बढ़ती जनसंख्या के कारण घटित हो रहा है, क्योंकि उसकी वजह से पर्यावरण का क्षरण हो रहा है।

11. वायु, जल, मृदा प्रदूषण के सतत विकास पर प्रभाव :

विश्व की सभी अर्थव्यवस्थाओं ने जरूरत से अधिक आर्थिक विकास करके अंधाधुंध तरीके से उत्पादन बढ़ाया। हमने बेहिसाब कारें बनाई पर जिनसे वो चलती हैं उन जैविक ईंधनों का असीमित दोहन किया। वायु व जल प्रदूषण फैलाने वाली फैक्टरियाँ लगाई, जिनसे पूरी दुनिया में प्रदूषणों के सभी प्रकारों का स्तर भयावह रूप से बढ़ा।

12. वर्तमान जागरूकता में मीडिया की भूमिका :

इस बिन्दु को इस पत्र में स्थान देने का मेरा उद्देश्य मीडिया की इस विषय के प्रति बरती जाने वाली उदासीनता है। विशेषकर भारतीय मीडिया की। थोड़े बहुत अच्छे एडियोरियल जो पर्यावरण संरक्षण पर मिलते हैं वे प्रिन्ट मीडिया में ही कुछ चुनिंदा अखबारों में मिलते हैं। हम जानते हैं कि इलैक्ट्रोनिक मीडिया का प्रभाव वर्तमान में अधिक पड़ता है तो यदि नित्य प्रति हर मीडिया चैनल इसे अपने प्रसारणों में स्थान देगा तो अधिक जागरूकता फैलेगी।

13. सतत आर्थिक विकास लक्ष्यों SDG 2030 पर कोविड-19 महामारी का दुष्प्रभाव :

हम सभी जानते हैं कि कोविड-19 महामारी, जिसमें कोरोना महामारी विश्व के 2 वर्ष लील गई, ने विश्व की अर्थव्यवस्थाओं व स्वास्थ्य को बुरी तरीके से प्रभावित किया। इस वजह से अध्ययन बता रहे हैं कि सतत विकास के 17 लक्ष्यों (SDG-2030) में से अधिकतर नकारात्मक रूप से प्रभावित हुए। यद्यपि पर्यावरण व पारिस्थितिक तंत्र का इस अवधि में उन्नयन देखा गया। कोविड-19 के SDG 2030 पर पड़े प्रभावों का उल्लेख U.N.O. तथा अन्य संस्थाओं द्वारा किए गए सर्वेक्षणों के आधार पर निम्नलिखित रहे :—

(अ) गरीबी से मुक्ति : 2015 से किए गए सतत प्रयासों के बावजूद कोरोना काल के 2019 से 2020 के दौरान विश्व के निम्न व मध्यम आय वर्ग वाले राष्ट्रों में गरीबी का स्तर बढ़ा। U.N.O. के आँकड़ों के अनुसार इस महामारी ने अनुमानित 120 मिलियन लोगों को घोर गरीबी में धकेल दिया है, जिससे अभी भी वे उबर नहीं पाए हैं, तो ये SDGs के लक्ष्य नम्बर-3 को नहीं प्राप्त करना हुआ।

(ब) अच्छा स्वास्थ्य और कल्याण : कोविड-19 में कम समय में ही तेजी से ही मृत्यु दर का बढ़ना SDG लक्ष्य -3 की प्राप्ति में रोड़ा बना है। यहाँ तक की विकसित देशों में यह आँकड़ा अधिक चिंताजनक रहा है। मात्र इतना ही नहीं विभिन्न देशों में किए गए U.N.O. एवं N.G.O. के द्वारा किए गए सर्वेक्षण बताते हैं कि जो बीमारी से ग्रस्त हुए उनमें और जिनका रोजगार, आय छिन गए उनमें अवसाद, मानसिक तनाव बढ़ा है जिनका दीर्घकालिक प्रभाव राष्ट्रों की उत्पादता पर पड़ेगा।

(स) गुणवत्तापूर्ण शिक्षा का मिलना : कोविड से सतत विकास लक्ष्यों में से ये SDG-4 भी हासिल नहीं हो पाया क्योंकि महामारी के कारण विश्व की शिक्षण संस्थाएँ लम्बे समय तक बंद रहीं। विकल्प के तौर पर ऑनलाइन शिक्षा को

Pollution and climate change

अपनाया गया। लेकिन गरीब राष्ट्रों में एन्ड्रायड मोबाईल, इंटरनेट जैसी सुविधाओं के अभाव ने इसे प्रत्येक बच्चे तक नहीं पहुँचने दिया। शिक्षकों से सम्पर्क जितना ऑफलाइन रहता है, उसमें बच्चे गुणवत्तापूर्ण शिक्षा ग्रहण करते हैं। कोविड़ –19 ने इस लक्ष्य की प्राप्ति को दूर की कौड़ी बना दिया।

(द) पूर्ण रोजगार और उत्पादक गतिविधियाँ : इन लक्ष्यों में से यह (SDG-8) 8वाँ लक्ष्य था, लम्बे लॉकडाउन के चलते पूरे विश्व में इसे हासिल करने में अब लम्बा समय लगेगा क्योंकि पूरा विश्व अब स्फीति मिश्रित मंदी का दौर देख रहा है। जिसे अर्थशास्त्र की भाषा में **Stagflation** कहते हैं।

(य) वैश्विक लक्ष्यों की प्राप्ति में साझेदारी : यह SGD 2030 का 17वाँ लक्ष्य है। जिसकी प्राप्ति में कोविड़–19 ने अवरोध पैदा किए। अन्तर्निर्भरता की वैश्वीकरण की भावना अब आत्मनिर्भरता वाली राष्ट्रवादी सोच में तबदील हो चुकी है। यहाँ तक कि कोविड़–19 महामारी से निजात पाने के प्रयासों में जो वैक्सीन निर्माण व वितरण की प्रक्रिया थी, वह व्यावसायिक दृष्टिकोण लिए रही और मानवीय पक्ष की अवहेलना हुई। परिणामस्वरूप गरीब देशों तक उसका वितरण नहीं हो पाया। इसे मेरे विचार में **U.N.O.** की एक विफलता के रूप में देखा जाना चाहिए।

14. क्या सतत विकास के विचार को हकीकत के धरातल पर उतारना संभव है ?

मेरे इस शोध–पत्र में बहुत से विचारणीय मुद्दों के अलावा एक प्रश्न का उत्तर खोजना है कि जिस विषय पर पूरी दुनिया में बहस छिड़ी है और मानवता को बचाने के लिए, इस पारिस्थितिकी तंत्र की धरोहर को बनाए रखने के लिए ये जरूरी भी है लेकिन क्या यह संभव भी है या नहीं ?

तो इसका उत्तर यही है कि यह संभव है यदि सारी दुनिया के राष्ट्र इसके लिए कृत संकल्प हों तथा ईमानदारी से अपना–अपना रोल निभाएं। ऐसा इसलिए क्योंकि दुनियां वो नहीं वरन् तीन ध्रुवों में बंटी है। विकसित तथा विकासशील के अतिरिक्त एक तीसरी दुनिया के देश भी हैं जो प्रति व्यक्ति आय व उपभोग की सीमांत मात्रा से कम प्राप्त कर रहे हैं। अतः उनके द्वारा सतत विकास बनाए रखने की कड़ी में बराबर का योगदान लेने की कल्पना एक अनाधिकृत प्रयास है जो कि विकसित राष्ट्र हर मंच पर करते आए हैं, चाहे वो पृथ्वी सम्मेलन हो, **U.N.O.** के पर्यावरण चिंता सम्बन्धी सम्मिट हो, **W.T.O.** का मंच हो अथवा पर्यावरणविदों द्वारा उठाए गए प्रश्नों का जबाव देना हो।

यदि विकसित राष्ट्र और भारत, चीन, ब्राजील जैसे विकासशील राष्ट्र अपने–अपने रूतबे या उपलब्धि से साधनों के हिसाब में योगदान दे तो सतत विकास लक्ष्यों **(S.D.G.-2030)** को प्राप्त करना संभव है।

एक उदाहरण द्वारा समझाना चाहूँगी जब स्वीडन की एक 14 वर्षीय बालिका ग्रेटा थनबर्ग अकेले पर्यावरण की चिंता के लिए आवाज बुलन्द कर सकती है और पूरी दुनिया तथा **U.N.O.** का ध्यान इस मुद्दे की ओर आकर्षित कर सकती है तो पूरे विश्व को बचाने के लिए हम कुछ न कुछ योगदान अवश्य दे सकते हैं।

Pollution and climate change

15. विकास को टिकाऊ अथवा धारणीय बनाने के लिए किए गए वैश्विक, सरकारी व संस्थानिक प्रयास :

जब हम बात करते हैं S.D.G.-2030 लक्ष्यों की जिनका मुख्य उद्देश्य ही समाप्ति की ओर जा रही धरती को समय रहते बचाना है तो इसका अभिप्राय यह भी कतई नहीं है कि 2015 से पहले इस दिशा में कोई प्रयास नहीं किए गए। 2015 से पूर्व उठाए गए कदमों की संक्षिप्त चर्चा किए बिना यह शोध-पत्र अधूरा प्रतीत होगा। ये प्रयास बिंदुवार निम्न प्रकार चले :

(1) 1992 में संयुक्त राष्ट्र संघ ने बढ़ती ग्लोबल वार्मिंग की समस्या पर सक्रिय कार्य करने के लिए **UNFCCC (United Nations Framework Convention on Climate Change)** कार्यक्रम की शुरूआत की। इसमें किए गए कार्य व शोध के आधार पर यह निष्कर्ष दिया गया कि विकसित राष्ट्रों ने अंधाधुंध औद्योगिकरण द्वारा पर्यावरण को अधिक नुकसान पहुँचाया है और ग्रीन हाऊस गैसेज को बड़ी मात्रा में पर्यावरण में छोड़ने के लिए वे ही अधिक जिम्मेदार हैं।

(2) 1997 में जापान के शहर क्योटों में संयुक्त राष्ट्र संघ ने ग्लोबल वार्मिंग पर चर्चा करने के लिए एक अधिवेशन रखा। लेकिन किसी भी राष्ट्र ने पृथ्वी को पहुँचाए जाने वाले नुकसान को भविष्य में कम करने या रोकने में पहल करने की हिम्मत नहीं दिखाई।

(3) आखिरकार समस्या की गम्भीरता को देखते हुए संयुक्त राज्य अमेरिका और 37 अन्य औद्योगिक विकसित राष्ट्रों ने अपने स्तर पर ग्रीन हाऊस गैसों के उत्सर्जन को 2012 तक कम करने के लिए सहमत हुए। लेकिन साथ ही अमेरिका जैसे देश इस बात पर अड़ गए कि विकासशील देश भी बहुत अधिक ग्रीन हाऊस गैसों का उत्सर्जन कर रहे हैं। जैसे भारत, चीन आदि तो इन्हें भी क्योटो संधि के अनुरूप हमारे साथ कार्य करें। लेकिन भारत और चीन का मसला था कि अभी तो विकास की दौड़ में वो शामिल हुए हैं। अतः जितना नुकसान विकसित देश पृथ्वी के पर्यावरण को पहुँचा रहे हैं। उसे सुधारने की पहल भी उन्हें ही करनी चाहिए।

कुछ देश चाहे वे विकसित हों या विकासशील इन मंचों पर दायित्वों को वहन करने की बात पर अड़ियल रवैया रखते आए हैं। चीन-अमेरिका विवाद, भारत-अमेरिका विवाद इन मंचों पर बहुत ख्याति प्राप्त विवाद रहे हैं। जिनका कोई नतीजा नहीं निकला। जब तक कि S.D.G.-2030 के 17 लक्ष्यों की प्राप्ति में एक साथ 163 देशों ने सहमति जताई और अब इसकी पैरवी U.N.O. द्वारा की जा रही है। साल-दर-साल इन 17 लक्ष्यों की प्राप्ति में कौनसा देश कितने पायदान पीछे गया था ? आगे आया इसकी माप की जाती है और रिपोर्ट प्रकाशित की जाती है।

Pollution and climate change

16. निष्कर्ष :

कोविड़–19 एक आकस्मिक वैश्विक प्राकृतिक आपदा थी, जिसने सभी देशों की अर्थव्यवस्थाओं को बुरी तरह पछाड़ा है। लेकिन अंत भला तो सब भला की तर्ज पर अब गाड़ी पुनः पटरी पर आ चुकी है। लेकिन इस मुश्किल दौर ने पूरी दुनिया को बहुत बड़ी सीख दे दी कि प्रकृति के ताण्डव के आगे हम सभी क्षणभंगुर हैं। हमारी आर्थिक प्रगति भी तभी तक स्थायी रह सकती है, जब हम प्रकृति के साथ तालमेल बनाकर चलेंगे।

विवरणिका एवं सन्दर्भ ग्रन्थ सूची :

1. Career Development Learning Goals 2022 Editors : Sarah O'shea Olivia Groves, Kylie Austin & Jodi Lamanna.
2. Environment and Sustainable Development : M.H. Fulekar ISBN-13 9788132211655
3. Climate and Energy Decoded : Tushar Choudhary.
4. Climate Restoration : The only future that will sustain the Human Race :- Peter Fiekowsky.
5. पर्यावरण और पारिस्थितिक अध्ययन, बी.पी.सिंहकृवसुन्धरा प्रकाशन
6. Newspapers :- The Hindu, The Times of India, Rajasthan Patrika, Dainik Bhaskar, Punjab Kesari etc.
7. हमारा सामान्य भविष्य, पर्यावरण और विकास पर विश्व आयोग, ऑक्सफोर्ड प्रैस, दिल्ली
8. बियॉण्ड ग्रोथ : द इकोनोमिक्स ऑफ सस्टेनेबल डेवलपमेंट, एच.ई. डेल
9. पर्यावरण भूगोल : सविन्द्र सिंह, वसुन्धरा प्रकाशन
10. द इम्पैक्ट ऑफ कोविड़–19 ऑन S.D.G. Indicator by ASR Journal
11. Role of Media in Achieving Millenium Development Goal : Y. Sidhgoura and S. Jagdish B.
12. विश्व आर्थिक और सामाजिक सर्वेक्षण
13. यू.एन.ओ. रिपोर्ट ऑन एस.डी.जी. – 2030, 2020, 2021, 2022
14. भारत सरकार की सतत विकास लक्ष्यों पर सर्वेक्षण रिपोर्ट 2021.22

सह–आचार्य, अर्थशास्त्र,
डॉ. भीमराव अम्बेडकर राजकीय महाविद्यालय, श्रीगंगानगर
email : kavitamadhuban2719@gmail.com

12. Integration of Logistic Innovative Approach for Solving Multi-Objective Transportation Problem

Jitendra Singh and Anju*

Abstract

Multi-objective transportation problem is a most important part of linear programming in optimization Engineering. This type of problems plays an important role in industrial field to maximize the profits and minimize the transport costs. Various transportation problems deals with single objective function but in real life situations many organizations wants to achieve in multiple tasks in objective form for making best transportation of goods. The optimization of Multi-objective transportation is harder task than optimization of single-objective transportation. In this paper, we proposed a new integration of logistic innovative approach to solving multi-objective transportation problem that is helpful as decision makers for handing. The effectiveness of this approach is illustrated by numerical example.

Keywords : Multi-Objective Transportation Problem; Linear Optimization; Transportation Problem; Excel Solver Software

Introduction

In the real life situations some of the transportation problem that include multiple, incommensurable and inconsistent objective functions is called multi objective transportation problem. The objective of transportation model is to determine the shipping schedules that minimize the total transportation costs while satisfying supply and demand limits. The practical applications of the transportation problem can be extended to other areas of operations, including inventory control, employment scheduling and personnel assignment. Generally, problem associated with supplying goods from various sources to numerous destinations, known as transportation problem, these transportation problems relate to single objective. However, in real-life instances, all organizations want to attain multiple objectives while making transportation of commodities. Lee et al. used goal-programming approach for making decision about attaining multiple goals and Zeleny generated non-dominated basic feasible solution for solving multi-objective linear programming problem. In order for solving the single-objective and multi-objective transportation problems, there are different methods/ approaches are accessible. The origin of transportation was first produced by F. L. Hitchcock in (1941)[1]. He proposed a study on the distribution of a product from numerous sources to various localities. This proposal assumed the first prominent contribution to the solution of transportation problem. A. Charnes and W.W. Cooper (1954)[2] developed the steppingstone method for explaining linear optimization calculations in transportation problems. This research has made available a diversity of tools, which can be brought to be bear on executive problems. O. Kirca and A. Satir (1990)[3] proposed a heuristic for obtaining and initial solution for the transportation problem. R. R. K. Shajma and K.D. Sharma (2000)[4] generated a new dual based procedure for the transportation problems. P. Pandian (2013)[5], proposed new multi-objective programming approach for fuzzy linear programming problems and suggest the level-sum method to find an optimal fuzzy solution for solving fuzzy linear programming problem that satisfying all constraints. M. Zangiabadi and H. R. Maleki (2013)[6] introduced a fuzzy goal programming technique to solve multi-objective transportation problems with some non-linear membership functions. S. K. Bharatiand and S. R. Singh (2014)[7] applied intuitionistic fuzzy optimization method and a comparative study for solving multi objective linear programming problems. A. J. Khan and D. K. Das (2014)[8] used fuzzy multi objective optimization with reference to multi objective transportation problem. R.G. Patel and P.H. Bhathawala (2016)[9] introduced an innovative approach for the optimizing of transportation problem. K. Bharathi and C. Vijayalakshmi (2016)[10] proposed evolutionary algorithms for optimization of multi-objective transportation problem. P. Anukokilaa, B. Radhakrishnan and M. Rajeshwari (2017)[11] applied goal programming approach for optimization of multi-objective transportation problem. T. Karthy and K .Ganesan (2018)[12] used genetic algorithm for solving multi objective transportation problem. Lakhveer Kaur, Madhuchanda Rakshit, & Sandeep Singh (2018)[13] proposed a new approach for optimizing multi-objective transportation problem. K. K. Jain, R. Bhardwaj and S. Choudhary (2019)[14] used lexicographic goal programming approach for solving multi-objective transportation problem. Khilendra Singh and Sanjeev Rajan (2019)[15] proposed a matrix maxima method with a pareto optimality criteria for solving multi-objective transportation problem. A. Anju (2019)[16] applied a new using ranking function and Russell's method to optimization of hexagonal intuitionistic fuzzy fractional transportation problem. In order to extend this work, we proposed new approach for solving multi-objective transportation problem, which is based on making best allocations in cell with minimum costs. The main benefit of this integration of logistic innovative approach is that the obtained efficient optimal solution which leads to a compromise other solutions preferred by different decision makers.

Preliminaries

Definition 1 [Balance Multi-Objective Transportation Problem]: A multi-objective transportation problem is said to be balance, if total supply equal to total demand i.e. $\sum_{i=1}^{m} a_i = \sum_{j=1}^{n} b_j$, Otherwise multi-objective transportation problem is said to be unbalance, in case of unbalance there are two cases arises such as $\sum_{i=1}^{m} a_i < \sum_{j=1}^{n} b_j$ and $\sum_{i=1}^{m} a_i > \sum_{j=1}^{n} b_j$.

Case I if $\sum_{i=1}^{m} a_i < \sum_{j=1}^{n} b_j$, then add a dummy source to the transportation problem with capacity $\sum_{j=1}^{n} b_j - \sum_{i=1}^{m} a_i$ in order to make it a balance problem.

Case II if $\sum_{i=1}^{m} a_i > \sum_{j=1}^{n} b_j$, then add a dummy destination to the transportation problem with demand $\sum_{i=1}^{m} a_i - \sum_{j=1}^{n} b_j$ in order to make it a balance problem.

Definition 2 [Initial Basic Feasible Solution of Multi-Objective Transportation Problem]: A set of non-negative values of the decision variables ($x_{ij} \geq 0$) that satisfies the total supply and total demand of multi-objective transportation problem is said to be initial basic feasible solution.

Definition 3 [Optimal Solution of Multi-Objective Transportation Problem]: initial basic feasible solution is said to be optimal solution of multi-objective transportation problem if it optimize the total multiple transportation costs.

Mathematical formulation of Multi-objective Transportation Model

The multi-objective transportation problem can be represented in the following table with m number of sources (Factories) and n number of destinations (Markets).

	Market-1	Market-2	....	Market-n	Supply
Factory - 1	C^1_{11} C^2_{11} .... C^k_{11}	C^1_{12} C^2_{12} .... C^k_{12}	....	C^1_{1n} C^2_{1n} .... C^k_{1n}	a_1
Factory - 2	C^1_{21} C^2_{21} .... C^k_{21}	C^1_{22} C^2_{22} .... C^k_{22}	....	C^1_{2n} C^2_{2n} .... C^k_{2n}	a_2
....	....	....	....	....	....
Factory - m	C^1_{m1} C^2_{m1} .... C^k_{m1}	C^1_{m2} C^2_{m2} .... C^k_{m2}	....	C^1_{mn} C^2_{mn} .... C^k_{mn}	a_m
Demand	b_1	b_2	....	b_n	$\sum_{j=1}^{n} b_j = \sum_{i=1}^{m} a_i$

Where C^k_{ij} represents costs to transport from Sources F_i to Destinations M_i, a_i represents maximum storage capacity of Sources F_i, b_j represents maximum demand of destination M_j and x_{ij} represents amount to be transport from Sources F_i to Destinations M_i

Pollution and climate change

The proposed approach proceeds as:
Step 1: Formulate the following multi-objective linear optimization model

$$Min.Z = \left\{ Z_1, Z_2, \ldots Z_k \;/\; Z_k = \sum_{i=1}^{m} \sum_{j=1}^{n} C^k{}_{ij} x_{ij} \quad \forall\; k(\geq 2) \in N \right\}$$

Subject to $\quad \sum_{j=1}^{n} x_{ij} \leq a_i, \quad \sum_{i=1}^{m} x_{ij} \geq b_j, \quad \sum_{j=1}^{n} b_j = \sum_{i=1}^{m} a_i, \; x_{ij} \geq 0$

Step 2: Construct single objectives linear optimization models from multi-objective linear optimization model of step I and solve these problems by taking one of the objective functions at a time.

Problem-1
$$MinZ_1 = C^1{}_{11}x_{11} + C^1{}_{12}x_{12} + C^1{}_{13}x_{13} + C^1{}_{14}x_{14} + C^1{}_{21}x_{21} + C^1{}_{22}x_{22} + C^1{}_{23}x_{23} + C^1{}_{24}x_{24} + C^1{}_{31}x_{31} + C^1{}_{32}x_{32} + C^1{}_{33}x_{33} + C^1{}_{34}x_{34} + \ldots$$

Subject to $\quad \sum_{j=1}^{n} x_{ij} \leq a_i, \quad \sum_{i=1}^{m} x_{ij} \geq b_j, \quad \sum_{j=1}^{n} b_j = \sum_{i=1}^{m} a_i, \; x_{ij} \geq 0$

Problem-2
$$MinZ_2 = C^2{}_{11}x_{11} + C^2{}_{12}x_{12} + C^2{}_{13}x_{13} + C^2{}_{14}x_{14} + C^2{}_{21}x_{21} + C^2{}_{22}x_{22} + C^2{}_{23}x_{23} + C^2{}_{24}x_{24} + C^2{}_{31}x_{31} + C^2{}_{32}x_{32} + C^2{}_{33}x_{33} + C^2{}_{34}x_{34} + \ldots$$

Subject to $\quad \sum_{j=1}^{n} x_{ij} \leq a_i, \quad \sum_{i=1}^{m} x_{ij} \geq b_j, \quad \sum_{j=1}^{n} b_j = \sum_{i=1}^{m} a_i, \; x_{ij} \geq 0$

Step 3: Apply following Logistic innovative approach for results
$$MinZ = (C^1{}_{11} + C^2{}_{11})x_{11} + (C^1{}_{12} + C^2{}_{12})x_{12} + (C^1{}_{13} + C^2{}_{13})x_{13} + (C^1{}_{14} + C^2{}_{14})x_{14} + (C^1{}_{21} + C^2{}_{21})x_{21} + (C^1{}_{22} + C^2{}_{22})x_{22} +$$
$$(C^1{}_{23} + C^2{}_{23})x_{23} + (C^1{}_{24} + C^2{}_{24})x_{24} + (C^1{}_{31} + C^2{}_{31})x_{31} + (C^1{}_{32} + C^2{}_{32})x_{32} + (C^1{}_{33} + C^2{}_{33})x_{33} + (C^1{}_{34} + C^2{}_{34})x_{34} + \ldots$$

Subject to $\quad \sum_{j=1}^{n} x_{ij} \leq a_i, \quad \sum_{i=1}^{m} x_{ij} \geq b_j, \quad \sum_{j=1}^{n} b_j = \sum_{i=1}^{m} a_i, \; Z_k > Z_{k-1}, x_{ij} \geq 0$

Step 4: solve this process until all demand and supply are satisfied.

Illustrative Example
Cement Company has three factories F_1, F_2 and F_3 whose monthly capacities are 35, 30 and 25 mini loading van respectively. Monthly requirement of warehouses W_1, W_2, W_3 and W_4 from each factories are 30, 26, 19 and 15 respectively. What would be transportation plan of the multi-objective transportation problem with following

minimum distribution table?

Where house → Factories ↓	W_1	W_2	W_3	W_4	Capacity
F_1	6 18	10 22	9 19	7 16	35
F_2	5 16	7 14	5 11	6 12	30
F_3	8 24	10 22	8 18	7 15	25
Demand	30	26	19	15	90

Step-I given multi-objective transformation problem converted into following multi-objective linear programming
$Min.Z = \{Z_1, Z_2\}$, where

$Z_1 = C^1_{11}x_{11} + C^1_{12}x_{12} + C^1_{13}x_{13} + C^1_{14}x_{14} + C^1_{21}x_{21} + C^1_{22}x_{22} + C^1_{23}x_{23} + C^1_{24}x_{24} + C^1_{31}x_{31} + C^1_{32}x_{32} + C^1_{33}x_{33} + C^1_{34}x_{34}$

$Z_2 = C^2_{11}x_{11} + C^2_{12}x_{12} + C^2_{13}x_{13} + C^2_{14}x_{14} + C^2_{21}x_{21} + C^2_{22}x_{22} + C^2_{23}x_{23} + C^2_{24}x_{24} + C^2_{31}x_{31} + C^2_{32}x_{32} + C^2_{33}x_{33} + C^2_{34}x_{34}$

Subject to $x_{11} + x_{12} + x_{13} + x_{14} \leq 35$, $x_{21} + x_{22} + x_{23} + x_{24} \leq 30$,

$x_{31} + x_{32} + x_{33} + x_{34} \leq 25$ $x_{11} + x_{21} + x_{31} \geq 30$, $x_{12} + x_{22} + x_{32} \geq 26$,

$x_{13} + x_{23} + x_{33} \geq 19$, $x_{14} + x_{24} + x_{34} \geq 15$, $x_{ij} \geq 0$,

Step -2 Construct single objectives linear optimization problem-1, problem -2 by taking one of the objective functions at a time.

Problem-1

$MinZ_1 = C^1_{11}x_{11} + C^1_{12}x_{12} + C^1_{13}x_{13} + C^1_{14}x_{14} + C^1_{21}x_{21} + C^1_{22}x_{22} + C^1_{23}x_{23} + C^1_{24}x_{24} + C^1_{31}x_{31} + C^1_{32}x_{32} + C^1_{33}x_{33} + C^1_{34}x_{34}$

Subject to $x_{11} + x_{12} + x_{13} + x_{14} \leq 35$, $x_{21} + x_{22} + x_{23} + x_{24} \leq 30$,

$x_{31} + x_{32} + x_{33} + x_{34} \leq 25$

$x_{11} + x_{21} + x_{31} \geq 30$, $x_{12} + x_{22} + x_{32} \geq 26$, $x_{13} + x_{23} + x_{33} \geq 19$,

$x_{14} + x_{24} + x_{34} \geq 15$, $x_{ij} \geq 0$,

Pollution and climate change

Computational solution of the problem -1 (Table-1)

	x_{11}	x_{12}	x_{13}	x_{14}	x_{21}	x_{22}	x_{23}	x_{24}	x_{31}	x_{32}	x_{33}	x_{34}	U	SIGN	A
	30	0	0	5	0	11	19	0	0	15	0	10			
C1	1	1	1	1	0	0	0	0	0	0	0	0	35	<=	35
C2	0	0	0	0	1	1	1	1	0	0	0	0	30	<=	30
C3	0	0	0	0	0	0	0	0	1	1	1	1	25	<=	25
C4	1	0	0	0	1	0	0	0	1	0	0	0	30	>=	30
C5	0	1	0	0	0	1	0	0	0	1	0	0	26	>=	26
C6	0	0	1	0	0	0	1	0	0	0	1	0	19	>=	19
C7	0	0	0	1	0	0	0	1	0	0	0	1	15	>=	15
OV	$6	$10	$9	$7	$5	$7	$5	$6	$8	$10	$8	$7			
Min Z_1= 607															

	W1	W2	W3	W4	Capacity
F1	30	0	0	5	35
F2	0	11	19	0	30
F3	0	15	0	10	25
Demand	30	26	19	15	

$F_1 \xrightarrow{30} W_1 \Rightarrow 6 \times 30 = 180,$ $\quad F_2 \xrightarrow{19} W_3 \Rightarrow 5 \times 19 = 095$

$F_1 \xrightarrow{5} W_4 \Rightarrow 7 \times 5 = 035,$ $\quad F_3 \xrightarrow{15} W_2 \Rightarrow 10 \times 15 = 150$

$F_2 \xrightarrow{11} W_2 \Rightarrow 7 \times 11 = 077,$ $\quad F_3 \xrightarrow{10} W_4 \Rightarrow 7 \times 10 = 070$

$$\text{total cost} = 607 \text{ Units}$$

Problem -2

$MinZ_2 = C^2_{11}x_{11} + C^2_{12}x_{12} + C^2_{13}x_{13} + C^2_{14}x_{14} + C^2_{21}x_{21} + C^2_{22}x_{22} + C^2_{23}x_{23} + C^2_{24}x_{24} + C^2_{31}x_{31} + C^2_{32}x_{32} + C^2_{33}x_{33} + C^2_{34}x_{34}$

Subject to $x_{11} + x_{12} + x_{13} + x_{14} \leq 35, \; x_{21} + x_{22} + x_{23} + x_{24} \leq 30,$

$x_{31} + x_{32} + x_{33} + x_{34} \leq 25$

$x_{11} + x_{21} + x_{31} \geq 30, \; x_{12} + x_{22} + x_{32} \geq 26, \; x_{13} + x_{23} + x_{33} \geq 19,$

$x_{14} + x_{24} + x_{34} \geq 15, \; x_{ij} \geq 0,$

Computational solution of the problem -1 (Table-1)

	x_{11}	x_{12}	x_{13}	x_{14}	x_{21}	x_{22}	x_{23}	x_{24}	x_{31}	x_{32}	x_{33}	x_{34}	U	SIGN	A
	30	5	0	0	0	21	9	0	0	0	10	15			
C1	1	1	1	1	0	0	0	0	0	0	0	0	35	<=	35
C2	0	0	0	0	1	1	1	1	0	0	0	0	30	<=	30
C3	0	0	0	0	0	0	0	0	1	1	1	1	25	<=	25
C4	1	0	0	0	1	0	0	0	1	0	0	0	30	>=	30
C5	0	1	0	0	0	1	0	0	0	1	0	0	26	>=	26
C6	0	0	1	0	0	0	1	0	0	0	1	0	19	>=	19
C7	0	0	0	1	0	0	0	1	0	0	0	1	15	>=	15
OV	$18	$22	$19	$16	$16	$14	$11	$12	$24	$22	$18	$15			
Min Z_2= 1448															

Pollution and climate change

	W1	W2	W3	W4	Capacity
F1	30	5	0	0	35
F2	0	21	9	0	30
F3	0	0	10	15	25
Demand	30	26	19	15	

$F_1 \xrightarrow{30} W_1 \Rightarrow 18 \times 30 = 540,$ $F_2 \xrightarrow{9} W_3 \Rightarrow 11 \times 09 = 099$

$F_1 \xrightarrow{5} W_2 \Rightarrow 22 \times 5 = 110,$ $F_3 \xrightarrow{10} W_3 \Rightarrow 18 \times 10 = 180$

$F_2 \xrightarrow{21} W_2 \Rightarrow 14 \times 21 = 294,$ $F_3 \xrightarrow{15} W_4 \Rightarrow 15 \times 15 = 225$

total $\cos t = 1448$ Units

Step 3: apply following logistic inventory approach for results

$MinZ = (C^1_{11}+C^2_{11})x_{11} + (C^1_{12}+C^2_{12})x_{12} + (C^1_{13}+C^2_{13})x_{13} + (C^1_{14}+C^2_{14})x_{14} + (C^1_{21}+C^2_{21})x_{21} + (C^1_{22}+C^2_{22})x_{22} + (C^1_{23}+C^2_{23})x_{23} + (C^1_{24}+C^2_{24})x_{24} + (C^1_{31}+C^2_{31})x_{31} + (C^1_{32}+C^2_{32})x_{32} + (C^1_{33}+C^2_{33})x_{33} + (C^1_{34}+C^2_{34})x_{34} + ...$

Subject to $x_{11} + x_{12} + x_{13} + x_{14} \leq 35,$ $x_{21} + x_{22} + x_{23} + x_{24} \leq 30,$

$x_{31} + x_{32} + x_{33} + x_{34} \leq 25$ $x_{11} + x_{21} + x_{31} \geq 30,$ $x_{12} + x_{22} + x_{32} \geq 26,$

$x_{13} + x_{23} + x_{33} \geq 19,$ $x_{14} + x_{24} + x_{34} \geq 15,$

$12x_{11} + 12x_{12} + 10x_{13} + 9x_{14} + 11x_{21} + 7x_{22} + 6x_{23} + 6x_{24} + 16x_{31} + 14x_{32} + 10x_3$

$x_{ij} \geq 0,$

Computational solution of the problem -3 (table-3)

	x_{11}	x_{12}	x_{13}	x_{14}	x_{21}	x_{22}	x_{23}	x_{24}	x_{31}	x_{32}	x_{33}	x_{34}	U	SIGN	A
	30	5	0	0	0	21	9	0	0	0	10	15			
C1	1	1	1	1	0	0	0	0	0	0	0	0	35	<=	35
C2	0	0	0	0	1	1	1	1	0	0	0	0	30	<=	30
C3	0	0	0	0	0	0	0	0	1	1	1	1	25	<=	25
C4	1	0	0	0	1	0	0	0	1	0	0	0	30	>=	30
C5	0	1	0	0	0	1	0	0	0	1	0	0	26	>=	26
C6	0	0	1	0	0	0	1	0	0	0	1	0	19	>=	19
C7	0	0	0	1	0	0	0	1	0	0	0	1	15	>=	15
C8	12	12	10	9	11	7	6	6	16	14	10	8	841	>=	0
OV	$24	$32	$28	$23	$21	$21	$16	$18	$32	$32	$26	$22			
Min Z= 2055				Min Z_1=607				Min Z_2=1448						

	W1	W2	W3	W4	Capacity
F1	30	5	0	0	35
F2	0	21	9	0	30
F3	0	0	10	15	25
Demand	30	26	19	15	

$F_1 \xrightarrow{30} W_1 \Rightarrow \begin{Bmatrix} 06 \\ 18 \end{Bmatrix} \times 30 = \begin{pmatrix} 180 \\ 540 \end{pmatrix}$ $F_2 \xrightarrow{9} W_3 \Rightarrow \begin{Bmatrix} 05 \\ 11 \end{Bmatrix} \times 09 = \begin{pmatrix} 045 \\ 099 \end{pmatrix}$

$F_1 \xrightarrow{5} W_2 \Rightarrow \begin{Bmatrix} 10 \\ 22 \end{Bmatrix} \times 5 = \begin{pmatrix} 050 \\ 110 \end{pmatrix}$ $F_3 \xrightarrow{10} W_3 \Rightarrow \begin{Bmatrix} 08 \\ 18 \end{Bmatrix} \times 10 = \begin{pmatrix} 080 \\ 180 \end{pmatrix}$

$F_2 \xrightarrow{21} W_2 \Rightarrow \begin{Bmatrix} 7 \\ 14 \end{Bmatrix} \times 21 = \begin{pmatrix} 147 \\ 294 \end{pmatrix}$ $F_3 \xrightarrow{15} W_4 \Rightarrow \begin{Bmatrix} 07 \\ 15 \end{Bmatrix} \times 15 = \begin{pmatrix} 105 \\ 225 \end{pmatrix}$

total $\cos t = \begin{pmatrix} 0607 \ Units \\ 1448 \ Units \end{pmatrix}$

Conclusion and future work

In this paper, we proposed integration of logistic innovaapproach for optimizing multi-objective transportation problem that can be used ideal deal of success for solving this type of problem within a short time. This approach will be very lucrative for those decisions who are dealing with supply chain and logistic issues in optimization techniques. This approach is efficient as compare with the other method and requires less time to solve multi-objective transportation problem. We therefore, hope that this approach may be used as an effective tool for solving this type of problem and hence time and labor may be saved. This approach can be used for assignment problems, travel salesperson problem and fuzzy based optimization techniques as future research studies.

Acknowledgement

We would like to thank Council of Scientific and industrial Research for providing a research fund to complete this work.

References

1. F.L. Hitchcock (1941). The Distribution of A Product from Several Sources To Numerous Localities, Journal Of Mathematical Physics 20 1941 224-230.
2. A. Charnes and W.W. Cooper (1954). The Stepping-Stone Method For Explaining Linear Programming Calculations In Transportation Problems, Management Science, 1(1), 49-69.
3. O. Kirca and A. Satir (1990). A Heuristic for Obtaining An Initial Solution For The Transportation Problem. Journal Of Operational Research Society, 41(9), 865-871.
4. RRK Shajma and Kd Sharma (2000). A New Dual Basedprocedure For The Transportation Problem. European Journal Of Operational Research,122, 611-624.
5. P. Pandian (2013), Multi-Objective Programming Approach For Fuzzy Linear Programming Problems, Applied Mathematical Sciences, Vol. 7, 2013, No. 37, 1811 – 1817.
6. M. Zangiabadi And H. R. Maleki (2013), Fuzzy Goal Programming Technique To Solve Multi-Objective

Transportation Problems With Some Non-Linear Membership Functions, Iranian Journal Of Fuzzy Systems Vol. 10, No. 1, (2013) Pp. 61-74.
7. S. K. Bharatiand And S. R. Singh(2014), Solving Multi Objective Linear Programming Problems Using Intuitionistic Fuzzy Optimization Method: A Comparative Study International Journal Of Modeling And Optimization, Vol. 4, No. 1, February 2014pp10-16.
8. A. J. Khan1 And D. K. Das (2014) Fuzzy Multi Objective Optimization: With Reference To Multi Objective Transportation Problem, Recent Research In Science And Technology 2014, 6(1): 274-282.
9. R.G. Patel And P.H. Bhathawala (2016). An Innovative Approach To Optimum Solution Of Transportation Problem. International Journal Of Innovative Research In Science, Engineering Technology 5(4): 5695-5700.
10. K. Bharathi And C. Vijayalakshmi (2016), Optimization Of Multi-Objective Transportation Problem Using Evolutionary Algorithms, Global Journal Of Pure And Applied Mathematics. Issn 0973-1768 Volume 12, Number 2 (2016), Pp. 1387-1396.
11. P. Anukokilaa, B. Radhakrishnan And M. Rajeshwari (2017), Multi-Objective Transportation Problem By Using Goal Programming Approach, International Journal Of Pure And Applied Mathematics Volume 117 No. 11 2017, 393-403.
12. T. Karthy And K .Ganesan (2018), Multi Objective Transportation Problem - Genetic Algorithm Approach, International Journal Of Pure And Applied Mathematics Volume 119 No. 9 2018, 343-350.
13. Lakhveer Kaur, Madhuchanda Rakshit, & Sandeep Singh (2018), A New Approach To Solve Multi-Objective Transportation Problem, Vol. 13, Issue 1 (June 2018), Pp. 150–159.
14. K. K. Jain, R. Bhardwaj, S. Choudhary (2019). A Multi-Objective Transportation Problem Solve By Lexicographic Goal Programming. Ijrte, Issn: 2277-3878, Vol. -7.
15. Khilendra Singh, Sanjeev Rajan (Sept. 2019), Matrix Maxima Method To Solve Multi-Objective Transportation Problem With

A Pareto Optimality Criteria, International Journal Of Innovative Technology And Exploring Engineering (Ijitee) Issn: 2278-3075, Volume-8 Issue-11pp 1929-1932.

16. A. Anju (2019). A New Method For Solving Hexagonal Intuitionistic Fuzzy Fractional Transportation Problem Using Ranking Function And Russell's Method. World Science News, Eissn: 2392-2192, Wsn 133 (2019), 234-247.

**Department of Mathematics,
R. R. College Alwar (Raj.) India.
email : jsambesh@gmail.com,
anju00955@gmail.com**

13. Impact of War on Biodiversity with particular reference to Critically Endangered Species
(An attempt to find solution to end up wars)

Prof. Krishan Kumar Sharma

Every day, we hear war news includes attack of missiles taking lot of lives, development of weapons to the end limit of nuclear weapons etc.. Lot of post war assessments and losses are done at the end of war. Assessment of damages caused by the war weapons on buildings, bridges, aero planes, helicopters, ships etc. but unfortunately, no assessment of loss to biodiversity, particularly the critically endangered species on earth. Buildings may be rebuilt, aero planes and ships may be manufactured again but would it be possible to regain those species at the critical status of IUCN. It took millions years for the earth to come in the present form supporting all living beings including humans with its assets. On the darker side of the scene, it takes seconds to destroy million years of evolutionary efforts of nature with high profile war weapons such as super sonic missiles, nuclear bombs, and other military weapons. A lot analysis and discussions on media about sufferers of wars such as death of people damage to buildings, Roads, bridges, aero planes, helicopters oil fields migration of people, refugees, sufferance etc. but no body in the world has ever raised issue on how much damage has been caused to the biodiversity in the wars in past few decades. I am going to share my views on this pertinent subject of great concern.

Theory of organic evolution suggest formation of various plants, animals and other microorganisms on earth through various mechanisms of evolution in millions of years. The **diversity** of the living world is staggering. More than 2 million existing **species** of organisms have been named and described; many more remain to be discovered—from 10 million to 30 million, according to some estimates. What is impressive is not just the numbers but also the incredible heterogeneity in size, shape, and way of life. From microscopic bacteria living in hot springs at temperatures near the **boiling** point of **water** to **fungi** and **algae** thriving on the ice

Pollution and climate change

masses of Antarctica and in saline pools at −23 °C (−9 °F). Living at the bed of oceans, mountains, desert and on the slopes of Mount Everest more than 6,000 metres (19,700 feet) above sea level. The virtually infinite variations on life are the fruit of the evolutionary process. All living creatures are related by descent from common ancestors. Humans and other mammals descend from shrew like creatures that lived more than 150 million years ago; mammals, birds, reptiles, amphibians, and fishes share as ancestors aquatic worms that lived 600 million years ago; and all plants and animals derive from bacteria-like microorganisms that originated more than 3 billion years ago. Biological evolution is a process of descent with modification. The beauty of environment is not concrete jungle but its Biodiversity. Biodiversity not only provide us food, medicine oxygen but a system in which we are connected with each other in some or different ways.

Biodiversity has been facing varieties of threats for its existence such as habitat loss, over exploitation, pollution, climate change and global warming, so many anthropogenic stresses. A lot has been written, discussed and published on these threats on biodiversity. But very little is known about habitat loss, species loss, genomic loss caused by the War weapons. How much damages during the war between different countries with high profile weapons like Missiles, Rockets, Bombs, Atom Bombs has damaged biodiversity has not become serious concern of Big powers and peace keeping organizations such as United Nations etc. Armed conflicts in war areas caused severe deforestation, soil erosion, and threats to wildlife. .Development and testing of high powered weapons including nuclear missiles, Chemical and Biological weapons on land ocean and air cause sever damage to the biodiversity, but no country pays attention on this serious issue .Rockets and missiles and bomb blast in a forest causing fire takes life of millions of organisms. Same thing happen when weapons in air leads to millions of avifauna in the affected area. A missile falling in an ocean in the fossil fuel area or war ship cause problem of oil spills and millions of marine organisms suffer of death and injuries. Loss of Herpetofauna in Russia-Ukraine war shall be discussed in detail.

When we have not yet completed documentation of more than 90 percent biodiversity, would it be wise to destroy them before that. In my opinion it would be impossible to regain genomic diversities of species resulting from million years of evolutionary processes and damaged in wars by any mean. When seminars and conferences would discuss, raise voice and send recommendations, a little hope to end up wars may come true.

Recommendation :

(i) International Court of Justice, one of the six principal organs of the United Nations, that gives advisory opinions on International legal issues, be given more powers to handle international conflicts in between countries particularly on issues related to damages caused by war to the biodiversity in general and critically endangered species in particular. Serious penalties be imposed on alleged nations after assessing damage to the biodiversity, considering it as a global property.

(ii) Reduction in the development and testing of war weapons on the Biodiversity rich areas. Total ban on mega diversity areas.

(iii) Vigorous awareness programmes be conducted as Conferences, seminars, workshops on the theme such as Impact of War on Biodiversity/damage to biodiversity in war.

(iv) Media plays important role in bringing issues before the people. Print and visual media should also look into damage caused by war weapons on biodiversity while covering ground reports of war. Social media should also report war victims of Biodiversity in their communications and viral messages.

(v) In the curriculum and curriculum frame work students be sensitize by introducing subject such as Impact of War on Biodiversity.

Former Vice Chancellor,
MDS University Ajmer (Rajasthan) India
email : kksmds@gmail.com

14. Impact of Climate Change on Indian Agriculture

Bharat Yadav*,
Dr Manju Yadav**

Abstract

"If agriculture goes wrong, nothing else will have a chance to go right."

- M. S. Swaminathan

"Climate change is a terrible problem that needs to be solved. It deserves to be a huge priority."

- Bill Gates, Founder of Microsoft

The projected pattern of the average weather of a region or the entire globe over a long period is significantly altered by climate change. The non-normal changes in the climate and their impacts are the subjects of this article. Climate change would alter the spatial and temporal distribution of environmental conditions and the intensity and frequency of weather and climate processes. The effects of climate change, a result of global warming, are now being felt everywhere. The effects of climate change include rising sea levels, an increase in the frequency of extreme weather, and adjustments to participation and growth rates.

Climate change has become a reality in India as well. Events brought on by climate change, such as floods, cyclones, droughts, heat waves, unseasonal rainfall patterns, etc., are affecting many regions of the nation. Highlighting the critical need for climate adaptation. (PIB, Feb. 2021) Extreme weather conditions negatively hurt agricultural output, endangering food security and nutrition while impeding the country's expansion and development. If climate change hurts agricultural production, livelihood and food security in India will be jeopardized. Because agriculture is the foundation of India's economy, climate change could lead to increased crop failure and pest outbreaks. As a result, future challenges will be more complex and difficult. This paper focuses on the variability of climate change and its probabilistic effects on agricultural productivity, as well as adaptation and mitigation strategies for managing the adverse effects of climate change on agricultural productivity, particularly in India.

Keywords : Climate Change, Agriculture, Temperature, Production, Impact, India

Introduction :

"Dreadful climate change events like floods, cyclones, droughts, heatwaves, unseasonal rainfall patterns etc., are affecting one or other part of the country, underlining the strong need for climate change adaptation. These extreme events severely affect farm production, threatening food and nutritional security and hampering the nation's growth and development. The demand for food is increasing considerably due to the increase in population. Hence, the top priority should be bringing harmony in the livelihoods of farming community by building resilience to the rural economy". (Shri N. S. Tomar, Minister of Agriculture and farmer's welfare, rural development and Panchayati Raj, Government of India)

Any significant long-term change in the expected patterns of average weather for a region or the entire planet over an extended period is defined as climate change. It includes non-normal climate variations as well as the consequences of these variations. Climate change has had a global impact, causing interconnected effects across society. The effects of climate change on various societal segments are interconnected. Food production and human health can be harmed by drought. Flooding can cause infrastructural and ecosystem damage as well as the spread of disease.

Since the climate of a region or nation impacts the nature and qualities of flora and crops, the agriculture sector is the one that is most vulnerable to climate change. Agriculture is negatively impacted by climate change, which results in a decline in productivity. This includes increased precipitation due to temperature changes and increased atmospheric CO_2 concentration, which may impact the crop's amount and quality.

Climate is the most important determinant of agricultural productivity, and it has a direct impact on food production. It, directly and indirectly, affects agriculture (Kang & Banga, 2013). Particularly in a growing nation like India, a country's economic and food security are significantly impacted by climate change. Climate change will affect all four dimensions of food security: food availability, food accessibility, food use, and food systems stability.

Climate Change and Agriculture :

Agriculture is the most vulnerable sector to climate change because a region's climate determines the nature and characteristics of vegetation and crops. Increased mean seasonal temperature can shorten the growing season of many crops and thus reduce final yield. Agriculture is a significant producer of greenhouse gases that contribute to the greenhouse effect, and climate change is projected to significantly impact future food availability and costs. Climate change severely impacts the agriculture industry, increasing the frequency and severity of natural disasters and threatening many people's lives and livelihoods. The impact of agricultural yield will be determined by the farmer's exposure to global environmental change and his or her ability to cope with and recover from global environmental change.

The production of food grains has been an essential part of India's mission to become self-sufficient in terms of the country's food requirements and reduce and eventually end its dependency on exports for food. This drive for self-sufficiency was one of the earliest motivations behind the Green Revolution in the 1960s. This period, led by M. Swaminathan, saw large-scale adoption of High Yield Variety seeds, modern farming equipment and chemical fertilizers. The result was that Punjab, where the revolution was first applied, began producing around 70 per cent of the nation's wheat and a 70 per cent increase in the farmers' income. The production of food grains in India, which was measly 50.82 million tons in 1950-51, has increased more than fivefold to 257.44 million tons in 2011-12.

The primary food grain crops in India are Rice and Wheat. Rice is a Kharif or summer season crop and is heavily water dependent. A country like India is heavily dependent on the monsoons, and any fluctuations can hurt its production. Wheat, on the other hand, is a Rabi crop and, while very adaptable, is affected by changes in temperature. Wheat production in India was deficient at the time of independence, 6.46 million metric tons but reached around 75.81

million tons in 2011-12. (Stats according to farmer.gov.in)

India's agricultural sector is susceptible to climate change. Food production is directly impacted by climate, which is the main factor in agricultural output. Climate change significantly impacts food production systems because it can cause disease and pest outbreaks, reduce harvests, and eventually threaten national food security. The effect on production is expected to vary depending on crop and location, also the magnitude of warming and the direction and magnitude of precipitation change (Adams et al., 1998).

Temperature increases affect the survival and distribution of pest populations, increase crop respiration rates, compress crop duration, the number of grains produced, and crop yield, inhibit sucrose assimilation in grains, hasten nutrient mineralization in soil, lower fertilizer use efficiency, and increase evaporation.

Higher temperatures often result in lower agricultural yields and more weed and insect growth. Recent research from the Indian Agricultural Research Institute suggests that every 1-degree Celsius increase in temperature during the growing stage could result in a loss of 4 to 5 M.T. in wheat yield in the future. Due to both the increase in temperature and the alteration in water supply, climate change harms the yields of irrigated crops. The length of many crops can be shortened by an increase in the mean seasonal temperature, resulting in lower ultimate yields.

Objectives of Study :
- To show the effects of climate change on food grain production in India through a review of literature on the topic
- To provide suggestions to the policymakers on tackling these effects on agricultural production.

Climate Change and Agriculture : A Literature Review

Global warming has been one of the significant problems which have affected humanity at large, and agriculture has not been averse to its effects. Global warming leads to climate change, leading to increased temperatures and a fluctuation in the rainfall levels in the country. The increasing temperatures will cause changes in precipitation, with uneven rainfalls causing dry areas to get drier and wet to get wetter. Thus, Rainfall changes, vital for a country like India, whose economy is mainly agriculture-based, can jeopardize the nation's economy. India's food grain production is called the gamble of the monsoon. Around 60 per cent of Indian agriculture is rainfed (Kumar et al., 2014). In contrast, almost 80 per cent of the rainfall is accounted for by the southeast monsoon, which occurs in the June to September window (Economic Survey, 2014). Furthermore, most Indian farmers, more than four-fifths, are small or middling level farmers and are thus much more vulnerable to changing patterns of monsoons. (Ranuzzi and Srivastava, 2012). A consistent rain pattern is usually required for agriculture; too much or too little rainfall, i.e., floods or drought conditions, can be damaging or even destructive to crops.

Global warming will cause a significant reduction in world agriculture this century. By 2080, it is anticipated that global agricultural productivity will have decreased by between 3 and 16 per cent. The 2080s will see an average 10 to 25% reduction in agricultural productivity in developing nations, many of which currently have average temperatures that are close to or above crop tolerance levels. Rich nations, whose average temperatures are normally lower, will see a considerably milder or even positive influence on production, which can range from a gain of 8% to a fall of 6%. Even greater decreases are faced by developing nations. India, for instance, may experience a 30 to 40% decline. (A. Mahato, 2014)

Government and international initiatives have predicted a worrying future if strategies aren't adopted for tackling the growing effects of global warming. According to the I.P.C.C. and a few other global studies, there will be a 10-40 per cent loss in crop production in India by 2080-2100 because of global warming (I.P.C.C., 2007a; Parry et al., 2004; Rosenzweig & Parry, 1994), despite the benefits of increased CO_2 concentrations.

Food and Agriculture Organization experts team concluded that every single degree increase in mean temperature would result in an annual wheat yield loss of around 6 million tons or USD 20 billion. (F.A.O., Food and Agriculture Organization, 2008; Swaminathan, 2012). Cline also stated that due to changes in climatic conditions, agricultural productivity, in general, could decline by 10-25 per cent by 2080, with rainfed agriculture yields declining by up to 50 per cent. (Cline, 2007)

A district-based study of Guiteras' on a 20-year data panel of over 200 districts has projected that a climate change over 2011-39 would lead to a decline of 4.5 to 9 per cent of agricultural yield, while in the long run (2070-99), output could decline by almost 25 per cent. (Guiteras, 2009)

There is a large literature of studies which have taken a whole nation-based approach to studying the impact of global warming. Most of these studies understandably deal with the effects on the food grain crops of Rice and Wheat.

Kaur and Hundals' findings from 2002 concluded that a 2-degree C rise in temperature resulted in an overall decline in both Kharif (grain/seed like rice, groundnut, and soybean) and Rabi crops (Wheat) by around 3-10 per cent and 29 per cent, respectively. (Prabhjyot-Kaur and Hundal, 2006). Another study by the pair in 2010 concluded that the temperature increase caused by the greenhouse effect hurts cereal and groundnut production. Although, the impact of temperature changes varies on crops and the direction of change. In the case of constant climatic variables, an increase in temperatures of 0.5, 1.0, 2.0, and 3.0 C above normal would advance wheat maturity by 3, 6, 12, and 17 days, respectively. (Prabhjyot-Kaur and Hundal, 2010)

Pandey et al.'s 2007 study discovered a progressive decline in yield of about 8 to 31 per cent when simulation wheat grain production under incremental units of maximum temperature (1-3 C) using the CERES-wheat model. (Pandey et al., 2007)

Ruchita Shah & Rohit Srivastava's study on the effects of global warming on Indian agriculture shows a funnel-shaped relationship between mean temperature and agricultural yield. Using the data between 1990-91 to 2012-13 of mean temperature and all data sets on food grain production, including Rabi and Kharif, Shah and Srivastava conclude that there is a linear relationship between the two at a lower temperature. (Shah & Srivastava, 2017)

Sharma recently brought attention to the effects of climate change on insects, crop protection, and food security. He emphasized that crop losses would likely increase due to changes in insect population and geographic range, which would have an impact on crop productivity and food security. (Sharma, 2013)

According to B. Venkateshwarlu, a former director at the International Central Research Institute for Dry Land Agriculture in Hyderabad, climate change affects all three aspects of food security—availability, access, and absorption. Climate change has a particularly negative impact on the poor. Since they lack the funds to purchase the food, they are unable to receive it. This will have long-term impacts on your health and absorption. It affects agriculture by between 4 and 9 per cent per year. India's G.D.P. is 15% the result of edge agriculture. According to estimates, climate change has caused a 1.5 per cent decline in G.D.P. By 2030, wheat and rice yields will likely have decreased by 6 to 10%. Kharif crops would be more impacted by rainfall variability, whereas the lowest temperature will more impact Rabi crops.

Many researchers have studied the state-wise and regional effects of Global warming and climate change-induced multi-faceted effects on the production of food crops of the respective regions. The major trend in the studies reviewed below has been that of decline.

According to Hundal et al., a one-to-three-degree Celsius increase in minimum temperature above normal resulted in a 3% decrease in rice productivity and a 10% decrease in wheat productivity in the state of Punjab. (Hundal et al., 2007)

Ahlawat, S. and Kaur, D. have studied the effect of fluctuating temperature and rainfall on the Wheat and Rice crops in the north-western states of Jammu and Kashmir, Himachal Pradesh, Haryana, and Punjab, also called the 'food bowl of India'. Working on data from 1901 to 2006, the paper concludes a positive correlation between rainfall and regional rice production. (Ahlawat & Kaur, 2007)

Samra and Singh argue that due to an increase in temperature between 3-6 C in the Indo-Gangetic plains, which is equivalent to nearly 1 C per day throughout the entire crop season, the wheat crop matured 10–20 days earlier and that the nation's wheat production fell by more than 4 million tons. (Samra and Singh, 2004)

Sahoo and Sridevi's study on climate change effects on Orissa's Foodgrains concludes that the summer rainfall, they conclude, is of utmost importance for the plantation of Kharif crop and readying the soil for cropping. The net result of the Kharif crop is also positive, but the autumn rains can cause problems if they lead to standing water. In case of increased rainfall, results are boosted with diminishing rate to a certain extent, after which it flattens. Further, Kharif crops are often destroyed because of floods in excess rains. Temperature conversely has a positive effect on foodgrains. (Sahoo & Sridevi. 2017)

Saseendaran, S N et al. have studied the effects of climate change on rice production in Kerela and have results show a continuous decline in the crop yield for up to 5 degrees increase in temperature, decrease in rainfall led to a yield loss of 8 per cent per 2mm/day. Further, the change in Kerela's temperatures from 1951 to 2007 has been studied by Rao, G.H.S.L.V. Prasada et al. (Rao, Prasada et al., 2008). The state's climate shift from wetness to dryness has seen changes in cropping patterns and has increased the chances of floods and droughts in the region. Similar declining effects have been found in the production of Rice in Tamil Nadu by Geetalakshmi et al. (Geetalakshmi et al., 2011) and Jowar production in Karnataka by Sushila and Ghasi (Sushila & Ghasi, 2009)

Suggested Important Adaptation and Mitigating Strategies to Minimize the Effects of Climate Change

While the previous section dealt with the relationship between and the impact of changing climatic conditions on the production of food grains, this section deals with strategies which should be the main areas of policy concern to mitigate the effects of global warming. The policy makers should:

(1) Establishing a "Green Research Fund" to advance research on adaptation, mitigation, and impact assessment.

(2) Advancing the widespread adoption of scientific and economical pricing policies for water, land, energy, and other natural resources; and

(3) Providing financial incentives for better land management.

Pollution and climate change

(4) Protecting the availability of food and income.
(5) Establishing seed banks in very erratic and variable settings, etc.
(6) More funding should be allocated by Indian governments to water storage and water use technology. To make better use of the water that is presently available, investments should be made in technology that enables micro irrigation and aquifer recharge.
(7) More funding should be allocated to initiatives focusing on land-use strategies and cultivars with climatically resistant features to maintain an adequate food supply.
(8) To create the necessary crop cultivars, a combination of conventional, molecular marker-assisted, mutational, and transgenic breeding procedures would be needed. Coordinated crop-based efforts need to get started to create climate-robust cultivars as soon as feasible.
(9) Under a climate change scenario, agronomic management of crops, such as sowing technique, can be a successful adaptation approach. Climate change has made bed planting of crops successful since it improves interrow cultivation and water use efficiency. It reduces waterlogging, weed management, fertilizer banding, stand establishment, and crop lodging while also requiring lower seeding rates.
(10) Zero-till systems are currently viewed as a feasible solution to tackle climate change. In irrigated areas, zero-tillage in wheat farming has successfully decreased the demand for water and other resources (e.g., fuel and pesticides).
(11) Technology is the main force behind adaptation to climate change. Awareness-raising and capacity-building among all stakeholders, from farmers to policymakers, are essential for the total adaptation to climate change.
(12) For people to adapt to climate change, community-driven programmers and village institution involvement are essential. Disseminating information about the harmful effects of climate change will help farmers become less vulnerable by enhancing their capacity for adaptation.

The only way to prepare our community, area, country, and societies for the effects of climate change is through adaptation and mitigation. In response to climate change, humans or natural ecosystems must adapt, which can somewhat hurt or distract from opportunities. In particular, it entails altering routine tasks due to a change in the climate but not entirely changing them; instead, it entails actively adjusting the existing habit. As a result, adoption risk management may need intricate governance procedures if technological advancement lowers output-per-unit input and emissions (GHGs into the atmosphere). India has very little room for further horizontal expansion to fulfil the rising need for food, fodder, fibre, fuel, and other items. Only vertical extension is possible within the scope. Agronomic management techniques could be beneficial in the fight against climate change.

Conclusion :

As agriculture is vulnerable to seasonal, yearly, and long-term climatic variations and short-term weather changes, there are numerous uncertainties with the emergence of risks from climate change. Agricultural productivity is greatly influenced by changes in meteorological parameters and other factors, including soil properties, cultivars, and pests (Pathak). Although there are climatic changes worldwide, developing nations like India, which have fewer resources than industrialized nations to deal with these adverse effects, are most severely affected by climate change. Therefore, it is challenging to ensure sustainable food security in emerging nations, particularly in India, due to the growing human population, rising demand for and intensity of resource usage, and increased per capita consumption (Rosenzweig & Parry, 1994).

In order to address the effects of climate change on Indian agriculture, the Indian Council of Agricultural Research (I.C.A.R.) launched the network project N.I.C.R.A. in 2011. According to research conducted as part of the N.I.C.R.A. project, climate change is anticipated to impact yields, notably those of rice, wheat, and maize. In the last three decades, there has been an increase in the frequency of extreme rainfall events and a rise in the mean temperature across India. The production of essential crops varies as a result of different years. Under the national innovation in resilient climate agriculture, the impact of climate change on Indian agriculture was investigated. The yields of irrigated and rainfed rice in India are predicted to decrease by 7 and 10 per cent, respectively, in 2050 and 2080.

Climate change's adverse effects on agriculture and its consequences on agriculture have serious ramifications that are expected to significantly impact food production and could jeopardize food security. As a result, special agricultural measures are needed to combat these effects. Careful resource management, including the management of soil, water, and biodiversity, will be necessary to deal with the effects of climate change on agriculture. India will need to take action at the international, regional, national, and local levels to address the effects of climate change on agricultural and food production.

References :
Adams, R.M., Hurd, B.H., Lenhart, S., Leary, N., 1998. Effects of global climate change on agriculture: an interpretative review. Climate Res. 11, 19–30
Ahlawat, Savita & Kaur, Dhian. (2015). Climate change and food production in North West India. Indian Journal Of Agricultural Research. 49. 10.18805/ijare.v49i6.6683.
Bushra, P. and Sharma, P.(2019)," A review of literature on climate change
Cline, W. R., 2007. Global Warming and Agriculture: Impact Estimates by Country. Peterson Institute for International Economics, Washington, DC
and its impacts on agricultural productivity", Journal of Public Affairs, Vol. 19, Issue
Datta, P. and All, (2022), "Climate change and Indian Agriculture: A systematic review of farmer's perception, adoption and transformation", Environmental Challenges, Vol. 8,
Economic Survey, 2013-14, Planning and Coordination Department, Directorate of Economics and Statistics, Government of India.

F.A.O. (Food and Agriculture Organization), 2008. Climate Change and Food Security: A Framework Document. F.A.O. of the United Nations, Rome, 107 pp

Hundal, S. S. (2007). Climatic variability and its impact on cereal productivity in Indian Punjab. Current Science, 506-512.

Guiteras, R. (2009). The impact of climate change on Indian agriculture. Manuscript, Department of Economics, University of Maryland, College Park, Maryland.

Geethalakshmi, V., Lakshmanan, A., Rajalakshmi, D., Jagannathan, R., Sridhar, G., Ramaraj, A. P., &Anbhazhagan, R. (2011). Climate change impact assessment and adaptation strategies to sustain rice production in the Cauvery basin of Tamil Nadu. Current Science, 342-347.

Hari, S., Khare, P., & Subramanian, A. Climate change and Indian agriculture

IPCC, 2007a. Contribution of working groups I, II, and III to the fourth assessment report of the intergovernmental panel on climate change. In: Pachauri, R.K., Reisinger, A. (Eds.), Climate Change 2007: Synthesis Report. I.P.C.C., Geneva, Switzerland

Kumar, A; Sharma, P; & Ambrammal, S.K.(2014), "Climatic Effects on Food Grain Productivity in India: A Crop Wise Analysis". Journal of Studies in Dynamics and Change, 1(1), 38-48

Kang, M.S., Banga, S.S., 2013. Global agriculture and climate change: a perspective. In: Kang, M.S., Banga, S.S. (Eds.), Combating Climate Change: An Agricultural Perspective. CRC Press, Boca Raton, FL, pp. 11–25

Mahato, A., (2014), Climate change and its impact on agriculture, International Journal of Scientific and Research Publications, Vol.4, Issue 4, ISSN 2250-3153

Parry, M.L., Rosenzweig, C., Iglesias, A., Livermore, M., Fischer, G., 2004. Effects of climate change on global food production under SRES emissions and socio-economic scenarios. Glob. Environ. Chang. 14, 53–67.

Pandey, V., Patel, H.R., Patel, V.J., 2007. Impact assessment of climate change on wheat yield in Gujarat using CERES-wheat model. J. Agrometeorol. 9, 149–157.

Prabhjyot-Kaur, Hundal, S.S., 2010. Global climate change vis-a`-vis crop productivity. In: Jha, M.K., Jha, M.K. (Eds.), Natural and Anthropogenic Disasters: Vulnerability, Preparedness and Mitigation. Capital Publishing Company and Springer, New Delhi and The Netherlands, pp. 413–431.

Prabhjyot-Kaur, Hundal, S.S., 2006. Effect of possible futuristic climate change scenarios on productivity of some Kharif and rabi crops in the central agroclimatic zone of Punjab. J. Agric. Phys. 6, 21–27.

Ranuzzi, A, & Srivastava, R. (2012), "Impact of climate change on agriculture and food security". I.C.R.I.E.R. Policy Series 16.

Rao, G.S.L.H.V.P., Ram Mohan, H.S., Gopakumar, C.S. and Krishnakumar, K.N. 2008. Climate change and cropping systems over Kerala in the humid tropics. Journal of Agrometeorology (Special issue, Part 2): 286-291.

Rosenzweig, C., Parry, M. Potential impact of climate change on world food supply. Nature **367,** 133–138 (1994). https://doi.org/10.1038/367133a0

Sahoo, D, Sridevi (2017), G. Impact of climate change on Food Grain Yield Across Seasons and Altitudes in Odisha. The Indian Economic Journal. 41. 387-399.

Saseendran, S. A., Singh, K. K., Rathore, L. S., Singh, S. V., & Sinha, S. K. (2000). Effects of climate change on rice production in the humid tropical climate of Kerala, India. Climatic Change, 44(4), 495-514.

Samra JS, Singh G (2004) Heatwave of March 2004: impact on agriculture. Indian Council of Agricultural Research, New Delhi

Shah, R & Srivastava, R (2017) Effect of Global Warming on Indian Agriculture Sustainability in Environment. 2(4). 366-378.

Srinivas Rao, Ravi Shankar Prasad and Trilochan Mohapatra, (2019) Climate Change and Indian Agriculture: Impacts, Coping Strategies, Programmes and Policies, I.C.A.R. Policy Paper.

Sushila, K., &Ghasi, R. (2009). Impact of global warming on the production of jowar in India. Agricultural Situation in India, 66(5), 253-256.

Swaminathan, M.S., 2012. Remember your Humanity: Pathway to Sustainable Food Security. New India Publishing Agency, New Delhi

Yoshida, S., Parao, P.T., 1976. Climatic influence on yield and yield components of low land rice in tropics. In: Proc. Symposium on Climate and Rice. International Rice Research Institute, Manila, Philippines, pp. 471–494.

Zhao, Chuang & Liu, Bing & Piao, Shilong & Wang, Xuhui & Lobell, David & Huang, Yao & Huang, Mengtian & Yao, Yitong & Bassu, Simona & Ciais, Philippe & Durand, Jean-Louis & Elliott, Joshua & Ewert, Frank & Janssens, Ivan & Li, Tao & Lin, Erda & Liu, Qiang & Martre, Pierre & Müller, Christoph & Asseng, Senthold. (2017). Temperature increase reduces global yields of major crops in four independent estimates. Proceedings of the National Academy of Sciences. 114. 201701762. 10.1073/pnas.1701762114.

*Research Scholar,
SOITS, IGNOU
Yadavbharat1502@gmail.com
*Associate Professor
Department of economics,
Govt College Ramgarh
Manjuyadav521@gmail.com

15. Indoor Air Quality : A Way to Sustainable Buildings

Dr. Bhawana Asnani

An Overview :

Indoor Air Quality (IAQ) refers to the air quality within and around buildings and structures, especially as it relates to the health and comfort of building occupants. Indoor air quality is important to human health because we spend more than 80 percent of our time indoors. Occupants of indoor environments may be exposed to a variety of pollutants originating from human activities or presence in the home, combustion for heating and cooking, consumer products, furnishings, building materials and outdoor air (https://www.epa.gov/indoor-air-quality-iaq/introduction-indoor-air-quality).

Indoor air quality depends on factors such as temperature, humidity, odours, air movement and ventilation, bioaerosols and volatile organic hydrocarbons. In some sealed buildings with the mechanical ventilation, there is a range of complaints, such as symptoms of eye and nose irritation, nasal congestion, sore throat, fatigue, malaise and headache, that are referred to as "sick building syndrome" (Burge and Feely, 1991). Besides many a biotic factors that can contribute to sick building syndrome, biological agents such as mold spores and bacteria play a significant role.

Composition of IAQ :

The quality of indoor air depends both on the quality of outdoor air and on the strength of emissions of indoor sources. In most inhabited spaces, there is a continuous exchange of air with the outside. Therefore, all contaminants of outdoor air are likely to be present indoors. The main pollutants in this category include Carbon monoxide, Nitrogen oxides, Sulphur oxides, Particulate matter, Hydrocarbons, Ozone and other photochemical oxidants and Lead.

These pollutants originate mostly from motor vehicles, factory emissions and other combustion processes. In the absence of indoor sources of these contaminants, their indoor concentrations will tend to be close to or lower than outdoor concentrations. Sulphur dioxide, ozone, soleplates, nitrates and lead concentrations, for example, are usually lower than those outside because of their reactivity and surface adsorptions (Kubba, 2017).

Generation of Indoor Air Pollution :

All indoor-generated airborne pollutants ultimately result from human activity. These pollutants can be categorized as follows:

- Those related to human activity or presence
- Those formed in combustion processes for heating and cooking
- Those derive from construction materials and furnishings.

Concentrations of pollutants in the first two categories tend to vary with time. Those in the third are likely to be more constant, provided that air exchange rates remain constant. Thus, control strategies for these three categories are likely to be different.

Modern building materials and furnishing can contribute indoor air pollutants. Adhesives, particle board, plywood, and carpets may emit significant quantities of VOCs and formaldehyde. Other potential sources of pollutants include insulation materials, including asbestos (no longer used in Australia) and lead based paints (no longer sold for domestic use but found in old buildings. The increasing use of water-based paints instead of solvent-based paints has reduced indoor emissions of VOCs and has also had advantages for outdoor air quality.The variety of contaminants resulting from human activity is very broad. Air fresheners, furniture waxes, polishes, cleansers, paints, pesticide formulations, fabric protectors, and deodorants are products frequently used in the home and are sources of various inorganic and organic chemicals. Many substances found in the workplace may also occur in the home as a result of hobby or craft activities(Petrovic, 2017).

Human metabolic activity itself influences air quality by reducing the concentration of oxygen and increasing levels of carbon dioxide. Respiration, perspiration and food preparation add water vapour as well as odour-producing substances to the indoor environment.

The air quality inside buildings is affected by many factors. In an effort to conserve energy, modern building design has favoured tighter structures with lower rates of ventilation. By contrast, in some areas of the world only natural ventilation is used; in other areas mechanical ventilation is common. Factors that can have a negative effect on health and comfort in buildings range from chemical and biological pollutants, to occupant perceptions of specific stresses such as temperature, humidity, artificial light, noise and vibration.

Moreover, indoor sources may lead to an accumulation of some compounds that are rarely present in the ambient air. The most important compounds in indoor air environments include SPM, SO_2, NO_2, CO, photochemical oxidants and lead. In developed countries, pollutant concentrations indoors are similar to those outdoors, with the ratio of indoor to outdoor concentration falling in the range 0.7-1.3 ppm. Concentrations of combustion products in indoor air can be substantially higher than those outdoors when heating and cooking appliances are used. This is particularly true in developing countries where ovens and braziers are used with imperfect kitchen and stove designs (https://www.niehs.nih.gov/health/topics/agents/indoor-air/index.cfm).

Indoor concentrations of air pollutants are influenced by outdoor levels, indoor sources, the rate of exchange between indoor and outdoor air, and the characteristics and furnishings of buildings. Indoor concentrations of air pollutants are subject to geographical, seasonal and diurnal variations.

Effects of Indoor Air Pollution on Human Health
One may feel the effects of exposure to an indoor pollutant immediately after exposure, or the problem may not show up until years later. Immediate effects include irritation of the eyes, nose, and throat; headaches; dizziness and fatigue. Age, preexisting conditions, and sensitivity to the pollutant can all affect whether a person reacts to a pollutant.Some health effects may show up shortly after a single exposure or repeated exposures to a pollutant. These include irritation of the eyes, nose, and throat, headaches, dizziness, and fatigue. Such immediate effects are usually short-term and treatable. Sometimes the treatment is simply eliminating the person's exposure to the source of the pollution, if it can be identified.

Other health effects may show up years after exposure or after repeated or long exposure. These effects can include central nervous system damage, chromosomal damage, and cancer. Health effects from exposure to combustion pollutants vary from very mild to lethal. Typical health effects are Headaches, Dizziness, Sleepiness, Nausea, Irritated eyes, Breathing difficulties Respiratory problems (i.e., coughing) etc.(https://www.epa.gov/indoor-air-quality-iaq/introduction-indoor-air-quality).

Inhalation of infectious microorganisms discharged by people and animals is a primary mechanism of contagion for most acute respiratory infections. In indoor environments characterized by reduced ventilation and increased use of untreated re-circulated air concentrations of microorganisms may increase.

Outdoor allergens, house dust mites, and moulds in indoor environments of high humidity can cause allergic asthma (reversible narrowing of lower airways), allergic rhino conjunctivitis in children and youth, and recurrent bouts of pneumonitis or milder attacks of breathlessness. The main acute effects of HCHO include odour perception and irritation of eyes, nose and throat. Discomfort, lacrimation, sneezing, coughing, nausea and dyspnea have also been observed, depending on the HCHO concentration. Health effects reported for VOC range from sensory irritation to behavioral, neurotoxic, hepatoxic and genotoxic effects. Concentrations at which identified health effects occur are usually much greater than those measured in indoor air. Exposure to mixtures of VOC may be an important cause of Sick Building Syndrome (SBS) Asbestos and other mineral fibers may be a cause of an increased incidence of lung cancer. Acute exposure to asbestos and glass fibers can cause severe skin irritation (Health Canada Factsheet).

More complex health effects are SBS and Building Related Illnesses (BRI). SBS is the occurrence of specific symptoms with unspecified etiology, and are experienced by people while working or living in a particular building, but which disappear after they leave it. Symptoms include mucous membrane, skin and eye irritation, chest tightness, fatigue, headache, malaise, lethargy, lack of concentration, odour annoyance and influenza symptoms. SBS usually cannot be attributed to excessive exposure to known contaminant or to a

defective ventilation system. A number of factors may be involved:
- Physical factors, including temperature, relative humidity, ventilation rate, artificial light, noise and vibration,
- Chemical factors, including environmental tobacco smoke, HCHO, VOC, pesticides, odorous compounds, CO, CO_2, NO_2 and O_3.
- Biological and psychological factors.

It is assumed that the interaction of several factors, involving different reaction mechanisms, cause the syndrome, but yet no clear evidence of any exposure-effect relationship (https://en.wikipedia.org/wiki/Indoor_air_quality).

BRI (Building Related Illness) is an illness related to indoor exposures to biological and chemical substances (e.g. fungi, bacteria, endotoxins, mycotoxins, radon, CO, HCHO). Lara, 2020 continues to state that some people working or living in a particular building experience it and it does not disappear after leaving it. Illnesses include respiratory tract infections and diseases, Legionnaires' disease, cardiovascular diseases and lung cancer.

Regulation of Iaq is The Solution :

IAQ in the built environment is controlled by an ever-increasing body of regulations and standards. Initially, standards for nondomestic buildings have aimed to control odor. During the development of ASHRAE Standard 62-1989 version, the committee set a minimum ventilation rate (Jones & Molina, 2017). Regulations of ventilation rates in houses often aim to control moisture and dilute the biproducts of combustion. However, these too are evolving to consider pollutant sources. ASHARAE defines *acceptable* IAQ as "air toward which a substantial majority of occupants express no dissatisfaction with respect to odor and sensory irritation and in which there are not likely to be contaminants at concentrations that are known to pose a health risk."

The WELL Building Standard (IWBI, 2016) focuses on the people in buildings, and identifies over 100 performance metrics, design strategies, and policies that can be implemented by building stakeholders to enhance the health and well-being of the occupants of buildings. The improvement of IAQ is one very important factor.

Source Control :

Usually the most effective way to improve indoor air quality is to eliminate individual sources of pollution or to reduce their emissions. Some sources, like those that contain asbestos, can be sealed or enclosed; others, like gas stoves, can be adjusted to decrease the amount of emissions. In many cases, source control is also a more cost-efficient approach to protecting indoor air quality than increasing ventilation because increasing ventilation can increase energy costs. Specific sources of indoor air pollution in your home are listed later in this section.

Improve Ventilation :

Another approach to lowering the concentrations of indoor air pollutants in your home is to increase the amount of outdoor air coming indoors. Opening windows and doors increases the ventilation rate. Bathroom or kitchen fans that exhaust outdoors remove contaminants directly from the room where the fan is located and also increase the outdoor air ventilation rate. If too little outdoor air enters a home, pollutants can accumulate to levels that can pose health and comfort problems. (https://www.cpsc.gov/Safety-Education/Safety-Guides/Home/The-Inside-Story-A-Guide-to-Indoor-Air-Quality).

Conclusion :

Indoor air quality has become an agenda using due to recent conservation efforts to reduce the cost of heating and cooking buildings, increased reliance on mechanical ventilation equipment, and growing awareness that air quality problems can impact the health, comfort and productivity of building occupants.In recent years, a growing body of scientific evidence has indicated the air people breathe inside their homes can be more seriously polluted with dangerous toxins than outdoor air in even the largest and most industrialized cities. The growing awareness of this issue is causing indoor air pollution to become a national health concern.

The occupants living or spending time in any kind of building, have a right to avail healthy indoor air. However the air in buildings can contain a range of pollutants that adversely affect the human body and mind. Exposure lowers physical and mental health, well-being, and productivity with social, political, and economic consequences. Providing plenty of fresh air is not always a solution because it has energy and carbon penalties. The improvement of IAQ is one very important factor that can be achieved by controlling the pollutant source and naturally ventilating the building.

References :
1. Burge, H. A. and Feely, J.C. 1991. Indoor Air Pollution and Infectious Diseases. Indoor Air Pollution, A Health Perspective, ed by Samet, J.M., Marbury, M.C., and Spengler, J.D. Baltimore MD: Johns Hopkins University Press, pp. 273-84.
2. http://www.hc-sc.gc.ca/hppb/tobaccoreduction/factssheet/indoors.htm.Health Canada factsheet. "Indoor air can become contaminated from many sources, but the most harmful and widespread contaminant of indoor air is tobacco smoke".
3. https://en.wikipedia.org/wiki/Indoor_air_quality
4. https://www.cpsc.gov/Safety-Education/Safety-Guides/Home/The-Inside-Story-A-Guide-to-Indoor-Air-Quality
5. https://www.epa.gov/indoor-air-quality-iaq/introduction-indoor-air-quality
6. https://www.niehs.nih.gov/health/topics/agents/indoor-air/index.cfm
7. Jones, B. & Molina, C. 2017. Sustainable Built Environment & Sustainable Manufacturing. Encyclopedia of Sustainable Technologies, pp 197-207.
8. Kubba, S. 2017. Indoor Environmental Quality. Handbook of Green Building Design & Construction (II Ed.). Butterworth-Heinemann. ISBN- 978-0-12-810433-0.
9. Lara, A.R. 2020. Building related Illness. MSD Manual-Consumer version. https://www.msdmanuals.com/en-in/home/lung-and-airway-disorders/environmental-lung-diseases/building-related-illnesses

10. Petrovic, E.K. 2017. A lack of recognition of potential health risks from building materials. Materials for a Healthy, Ecological and Sustainable Built Environment. Woodhead Publishing House. ISBN- 978-0-08-100707-5.

Assistant Professor,
Junagadh Agricultural University, Junagadh,
Gujarat
Website : www. bhawanaasnani.com,
email : bhawana_asnani@yahoo.com

16. Physico-Chemical Characterization of Farmland Soil in Rajuri Village of Rahata Taluka, Ahmednagar District, Maharashtra (India)

Vikhe A.S, Vikhe P.S, Kharde H.S,

Abstract

Physico-chemical properties of soil from different land were analyzed. Water holding capacity, available K, Ca, Li, Na, pH, electrical conductance, soil porosity, alkalinity, chloride, were studied. Soil samples were collected from the Ahmednagar district, Tal: Rahata, of Rajuri village, Maharashtra [India]. In the present study leads us to the conclusion of nutrients quantity of farmland soil. The main finding is that the parameters studied showed the differences in micronutrient content from one farmland to another in the same village. Analysis of micronutrients and physical parameters from different farmland soil will help the farmers to understand amount of fertilizer to be required to improve soil fertility and economic production of crops.

Keywords : Physico-chemical Parameters, Soil Fertility, Micronutrient, Flame Photometry, Rajuri.

Introduction :

The soil is one of the most essential top layers of earth surface. Soil is one of the best medium for plant growth and development. The physical properties of soil depend upon element shape, structure, size, organic matter and mineral composition. Knowledge of physical and chemical properties of soil helps in managing resources in soil [1]. Agriculture in India has occupied a diverse position in the cultivation of various crops [sugarcane, wheat, soya bean, cotton, black gram]. The importance of soil micronutrients in cultivation is carried out by studying various physical and chemical properties that help farmers to understand the fertility of soil to improve crop production as the properties of soil goes on changing due to anthropological activities. study of physiochemical parameter of soil in Rajura Bazaar in Amravati District of Maharashtra [India] was reported [2]. The second study was made in the Sangamner area, Ahmednagar District Maharashtra, [India]. Evaluating soil fertility [3]. Show moderate amount of soil micronutrients. Mineral elements in soil along with citrus fruit samples were studies that will be useful for some patients method used was a simple instrument Flame Atomic Absorption Spectrophotometer Different methods are used to analyze nutrients in soil [4, 5]. Soil type is red and black soil in adori region, the nitrogen level is medium, pH level is low of metal ions like K, Mg, Ca, and Iron in medium range. Farm level selected from different location consists of sugarcane cultivation land along with facilities of drip irrigation [7, 9]. Organic matter and physical chemical parameters observed like impact of crop on soil qualities to improvement in economic growth of farmer. General suggestion where given to farmers for shortage of micronutrient like urea for shortage of nitrogen, ferrous sulphate fertilizer for shortage of phosphorous. Plant nutrients like chloride, calcium, potassium, are available to plant at range of PH between 6.5to7.5 the soil studied was all in nature in pravaranagar area [13]. Different cropping patterns, soil micronutrients were studied in gogalgaon village in Ahmednagar district. Were sugarcane crop is difficult to cultivate as compare to tomato crop is completed in eight months [14]. Present study is aimed to find out the differences, variations in physical and chemical properties of soil.

Material Method :

Collection of soil sample was collected from the structure the surface of farm land soil from rajuri village in Tal: Rahata, Ahmednagar district, in the month of March 2022. Selected five representative soil samples were collected using the "V" shaped incision method. Tools like an auger and spade are used to dig a "V" shape incision in the sample with the depth of 15 cm. Soil sample dried, mixed and quarantine process was used. For storage of samples, polyether bags are used and labeled with information about the farmer. At the time of analysis, samples are ground and sieved, dried at 50^0 C, and passed through a 12 mesh screen. pH of soil was determined using pH meter, Electrical conductivity was determined using conductometer, K, Ca, Na, Li were determined using flame photometry instrument. Alkalinity, soil porosity and chloride content were determined using chemical method.

Result Discussion :

The physical and chemical parameters of farmland soil samples of Rajuri village studied are given in Table no.1 and all soil parameters variations are shown through graphical representation.

Table no. 1: Physico-chemical parameters of Rajuri village from different farm soil.

Sample no.	Physical parameter						Chemical Parameters			
	PH	E.C. ds/m	Soil moisture content(%)	Soil Porosity (%)	Alkalinity mg/L	Chloride mg/L	K (kg/ha)	Ca (kg/ha)	Li (kg/ha)	Na (mg/L)
1	7.2	0.320	4.24	40	123.7	193.3	131.6	474.16	116.4	1.76
2	6.8	0.143	2.90	52	135.4	155.6	154	377.26	122.08	0.704
3	6.5	0.225	1.40	34	225.2	323.4	117.6	423.36	224	1.408
4	7.1	0.344	7.04	55	275.6	410.1	106.4	383.84	337.12	1.584
5	6.2	0.454	5.62	66	323.4	163.2	126	286.94	226.24	1.496
Average Value	6.76	0.252	4.24	49.4	216.66	249.12	127.12	389.11	205.16	1.390

Figure no: 1

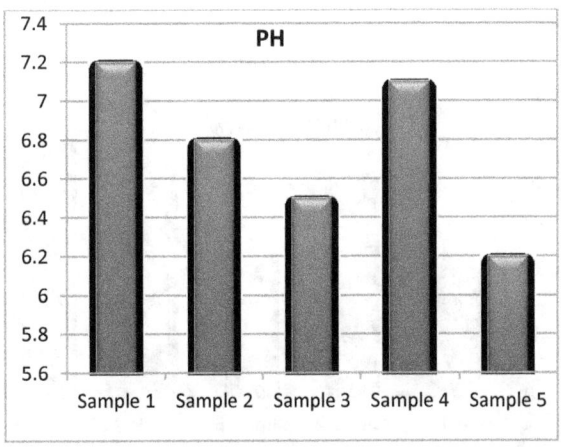

Figure no: 2

Pollution and climate change

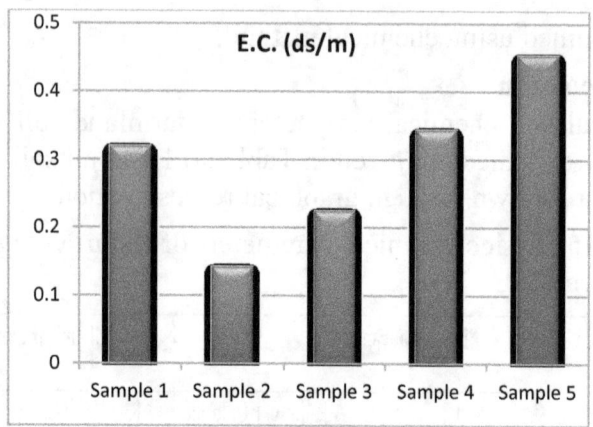

Figure no: 3

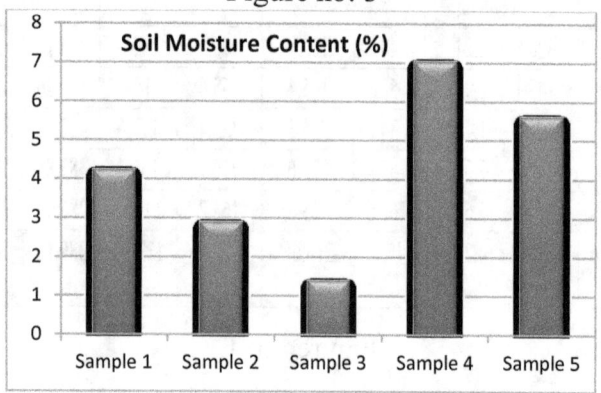

Figure no: 4

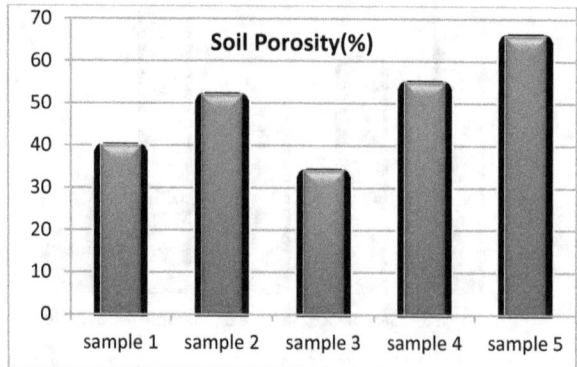

Figure no: 5

Pollution and climate change

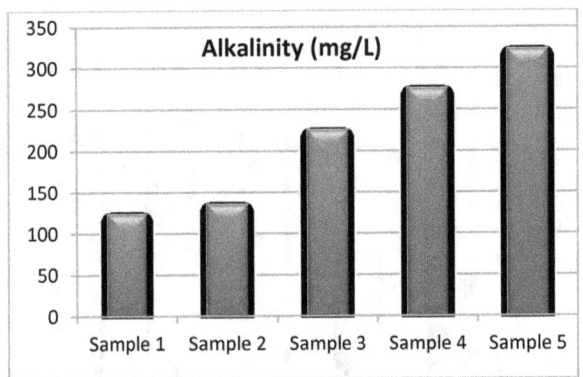

Figure no: 6

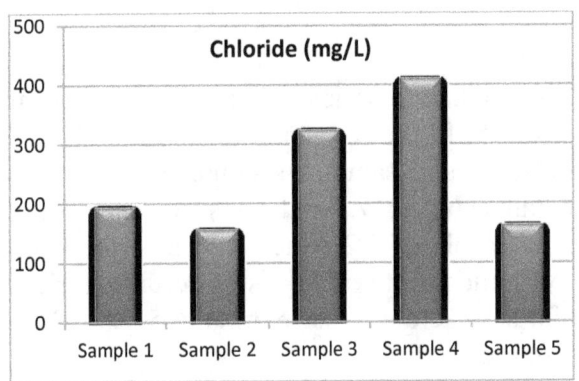

Figure no: 7

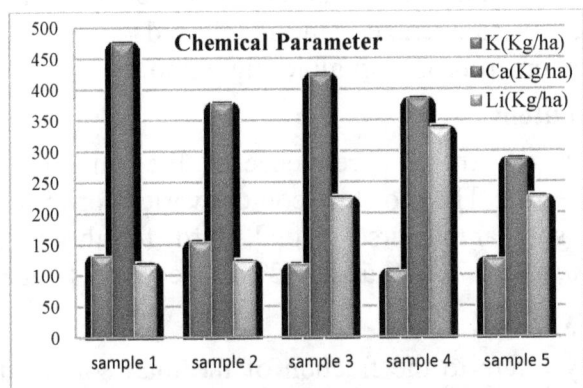

Figure no: 8

Pollution and climate change

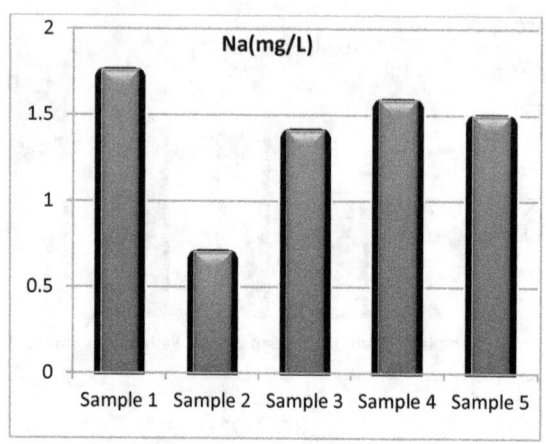

PH :

Soil prefers either alkaline or acidic condition. The normal PH range of soil is 6.5 to 7.5. In the soil sample first, the pH is 7.2 which is normal, and also in soil samples second, third, and fourth the pH range lies in range 6.5 to 7.5 indicates normal soil. Soil sample number fifth has a pH of 6.2 which indicates the acidic soil. At a higher pH, soil builds up toxic levels of certain nutrients. The limit of PH value for soil Acidic < 6.5, Normal 6.5-7.8, Alkaline 7.8- 8.5, Alkali > 8.5 [figure no.1].

Electrical Conductance :

The Electrical conductance in soil is in between 0 to 1 this indicates a non saline soil as average value is 0.252 ds/m. This type of soil salinity effect on crops is negligible [figure no.2].

Moisture Content :

The moisture content is a reference to the amount of moisture present in the soil. The moisture content varies concerning the type of soil. The soil sample first, second, third, fourth and fifth shows 4.24%, 2.9%, 1.4%, 7.04%, and 5.62% respectively [figure no.3].

Soil Porosity :

Soil porosity refers to the fraction of the total soil volume that is taken up by the pore space. The soil sample first has 40% porosity, the second has 52% porosity, the third sample has 34% porosity, the fourth has 55% porosity and the last one has 66% soil porosity [figure no.4].

Alkalinity :

Soil Alkalinity determines accumulation of soluble salts in soil. Presence of sodium carbonate and sodium bicarbonate indicates alkalinity, clay soils with high PH > 8.5 has poor soil structure and lower filtration capacity. The Average value of alkalinity in the selected soil sample is 216.66 mg/L [figure no.5].

Chloride :

Chloride in the soil is an essential nutrient present as chloride anion that maintains charge balance across a plant membranes act as osmoregulator. It is available through rain water, dust, air pollution, fertilizers. The average value of chloride in the selected soil sample is 249.12 mg/L [figure no.6].

Potassium :

Potassium is a critical nutrient that plants absorb from the soil and fertilizer. It increases disease resistance, helps stalks to grow upright and study, and improves drought tolerance. The observed amount of potassium in collected soil sample are 131.6 kg/ha, 154 kg/ha, 117.6 kg/ha, 106.4 kg/ha, 126 kg/ha respectively [figure no.7].

Calcium :

Calcium contributes to soil fertility by helping maintain flocculated clay and therefore good aeration. The observed amount of calcium in the soil sample are 474.16 kg/ha, 377.26 kg/ha, 423.36 kg/ha, 383.84 kg/ha, and 286.94 kg/ha [figure no.7].

Lithium :

Lithium stimulates or reduced plant growth depending on its concentration. The lithium in soil sample is found to be 116.4 kg/ha, 122.08 kg/ha, 224 kg/ha, 337.12 kg/ha, and 226.24 kg/ha respectively [figure no.7].

Sodium :

Sodium causes the soil pores to close up and make wetting the soil and water filtration difficult causing permeability problems. The sodium in the soil sample are found is 1.76mg/L, 0.704 mg/L, 1.408

mg/L, 1.584 mg/L, and 1.496 mg/L, respectively [figure no.8].

Conclusion :

In the present study, we analyze the physical and chemical parameters of soil samples from different localities of seasonal crops from Rajuri village. All the soil parameter such as PH, soil porosity, soil moisture content, alkalinity, chloride, electrical conductance, potassium, calcium, lithium, and sodium, is in various ranges. This study gives us information about the nature of the soil and presents nutrients in the soil. Physical chemical parameter of soil is significant in agriculture for growth of plant as well as soil management. This analysis is helpful to farmers of that area to arrange the amount as well as which fertilizers and nutrients are needed for soil to increase the percentage yield of crops. The present study indicates that the given soil is more suitable for a crop like sugarcane, wheat, cotton, soya bean, onion, and tomato fruit plants like grapes, bananas, pomegranate, etc.

References :
1. Tewari, G., Khati D, Rana, L., (2016) Assesment of Physicochemical properties of soils from different land use systems in Uttarakhand, India, *J. Chem. Eng. Chem. Res.*, (3), 1114-1118.
2. Ganorkar, P.R & Chinchmalatpure, P. G. (2013) Physicochemical Assessment of Soil in Rajura Bazar in Amravati District of Maharashtra (India), *International Journal of Chemical, Enviromental and Pharmaceutical research*, (4), 46-49.
3. Deshmukh, K. K (2012) Evaluation of soil Fertility Status from Sangamner area, Ahmednager districtof Maharashatra (India) *Rasayan J. Chem.*, (5), 398-406.
4. Khalilollah Taheri, Nahid Rahneshan,(2014) Determination of Mineral Elements in Soil, Citrus Limunum Juice and Citrus Limetta juice samples by Flame Photometry and Atomic Absorption Spectrometry Methods, *Iran J. Anal Chem* (1) 121-125.
5. Sonikajha, Suneetha V., (2015) Nutrient analysis of soil samples

from various places, *Journal of Chemical and Pharmaceutical Research*, 7(3), 291-293,
6. Ghare, P. M. Kumbhar, A. P. (2021) Study on Physico-Chemical Parameters of Soil Sample. *International Advanced Research Journal in Science, Engineering and Technology*, (8), 171-187.
7. Dhandwate, S. C. (2022) Analysis of soil samples for its physic-chemical parameters from sangamner city, *GSC Biological and Pharmaceutical Sciences*, https://doi.org/10.30574/gscbps.2020.12.2.0243 12(02), 123-128.
8. Ismail, M. Umamahesh, M. (2018) Analysis of soil samples for its physico chemical parameters in adoni region, kurnool district, (A.P), *Journal of Emerging Technologies and Innovative Research*, (5), 776-778.
9. Marshal, S. (2016) Analysis of soil samples for its physico-chemical parameters from abohar city, *The Pharma Innovation Journal*, 5(11), 37-39.
10. Sonawane, V.R, Gadekar D.J, (2020) Analysis of chemical properties of soil under sugarcane crop: Case study of khandala, shrirampur, Ahmednagar District, Maharashtra state, India,. 30(68), 6522-6549.
11. Ghatge, N., Rasal, O. (2013) Climate change: Causes, consequences & coping strategies, *International E-publication*.1-245.
12. Deshmukh, K. K., (2015) Study of Boron in Ground Water from Sangamner Area, Ahmednagar District, Maharashtra, *India Research Journal of Recent Sciences*, (4), 283-290.
13. Kamble, N.P., Kurhe, A.R., Pondhe, G.M., Gaikwad, V.B., (2013) Soil Nutrient Analysis and their relationship with special Reference to pH in Pravaranagar Area, District Ahmednagar, Maharashtra, India, *International Journal of Scientific & Techonology Research*, 3(2), 216-218.
14. Raut, R., Harale, P., Kurhe, A., (2022) Studies on Soil quality parameters in relations to cropping patterns, micronutrients, and pH from gogalgaon area in Ahmednagar District of Maharashtra, India. *International Journal for Innovative Research in Multidisciplinary Field*, 9(6), 210-219.

15. Achazi, R.K. (2002) Invertebrates in risk assessment development of a test battery and short term biotests for ecological risk assessment of soil, *Journal of Soil & Sediments,* (2), 174-178.
16. Chaudhari, K.G. (2013) Studies of the physicochemical parameter of soil sample, *Advances in Applied Science Research,* 4(6), 246-248.

Department of Analytical Chemistry,
Padmashri Vikhe Patil College of Arts, Science and Commerce,
Pravaranagar, Ahmednagar, Maharashtra, (India)
Department of Botany,
Padmashri Vikhe Patil College of Arts, Science and Commerce,
Pravaranagar, Ahmednagar, Maharashtra, (India)

17. House Plants Combating Indoor Pollution and Relieving Sick Building Syndrome

Madhulika Parmar

Abstract

In most of the urban dwellers, there is dearth of greenery and open spaces resulting in continuous confinement of people indoors without any access to greenery leading to Sick Building Syndrome. The increased energy efficiency of newer buildings with substantially reduced air exchange rates often results in an increase in the concentration of indoor air pollutants. NASA studies have found that indoor air pollution average 2-5 times higher than outdoors. The average person spends nearly 90% of his time inside, predominantly in homes, offices and schools and thus is refrained from exposure to a natural environment that is physically and mentally healthy. So, one has a smart choice to create a personal breathing zone by growing plants indoor. Researchers show that the air in plant-filled green rooms contains 50-60% fewer airborne moulds and bacteria compared to its equivalent. House planting can serve as the most efficient way to decrease the degree of SBSsymptoms as well as to regulate indoor air quality. Further regardless of the design and style, house plants blend seamlessly in every indoor setting. In recent years, house plants have become increasingly popular for interior decoration as well.

This review highlights the hazardous impacts of indoor air pollution to which most of the urban citizens are exposed, degrading their quality of life by making them physically and mentally disturbed which has been linked to Sick Building Syndrome. This paper also draws our attention towards a way of getting rid of this problem and the solution is to GO GREEN AND NATURAL. The benefits of the indoor plants are of great importance for indoor environment quality. Indoor plants help in purifying air, deter illness, ease tension, create a relaxed and happy atmosphere leading to work better through improved concentration, enhanced creativity, increased productivity and enriched overall well-being.

Keywords : House plants, indoor air pollution, indoor air quality, sick building syndrome, indoor plants.

Introduction

The unexplainable beauty of different formal and informal types of garden shows the love of man towards nature since time immemorial. House plants have been known as a part of human existence since 10,000 B.C. Nature did not create indoor plants, humans did this by taking plants that thrive in nature and bringing them indoors, where these plants have become an integral part of our lives. Usually indoor plants are chosen for their ability to tolerate particular indoor conditions such as low light, high temperature and low humidity. But, nowadays rapid urbanization, increase in vertical size of buildings, increase pollution from vehicles and factories, shrinkage of green cover makes it necessary to have greener environment both indoors and outdoors.

People today spend 80-90% of their time indoors and are thus alienating themselves from nature and limiting the opportunities to get exposed to the nature. So, in regard to this confinement of people within the four walls of the buildings, people have an increased concern about indoor air quality (Lim *et al.*, 2006). The indoor air in urban areas is reported to be much more polluted than outdoors (Yang *et al.*, 2004). It was reported that over 300 volatile organic compounds such asbenzene, toluene, ethylbenzene, and xylene (BTEX)(Craighead, 1995; Sullivan *et al.*, 2001; Wolkoff, 2002) and toxic chemical substances like formaldehyde have been identified as indoor contaminants in addition to dust and inorganic gases. These volatile organic compounds (VOCs) are regarded as the most important pollutants of indoor air. The indoor air pollutants originate from both living and inanimate objects within the dwelling and in some cases, diffuse into the building from the exterior and collectively, these pollutants result in a significant reduction in indoor air quality affecting the health, well-being and quality of life of those exposed to it. Hence, indoor air quality of any building is a basic determinant of healthy, overall well-being, comfort and productivity of the respondents. A study conducted by EPA states that indoor air pollution is among the top five environmental health risks. In 2009, the World Health Organization (WHO) prepared a report on 'Global Health Risks: Mortality and Burden of Disease Attributable to Selected Major Risks' stating that indoor air pollution is responsible for 2.7% of the global burden of disease. In 2012, indoor air pollution was linked to 4.3 million deaths globally, compared to 3.7 million outdoor air pollution. The deterioration of indoor air quality can result in "multiple chemical sensitivity, new house syndrome and sick building syndrome" (Shinohara *et al.*, 2004).

Sick Building Syndrome-A Serious Health Issue Affecting People's Well-Being

Sick Building Syndrome is described as a situation in which building occupants experience acute health problems and comfort effects that appear to be linked to time spent in a building, but no specific illness or cause can be identified. The complaints may be localized in particular room or zone or may be widespread throughout the building. The symptoms may include headache, dizziness, anxiety, insomnia, nausea, aches and pains, fatigue, poor concentration, shortage of breath or chest tightness, eye and throat irritation, etc. According to a World Health Organization working group (WHO, 1983), SBS symptom scan be divided into symptoms related to the mucous membranes (i.e. the eyes, nose and throat),dry skin, headache and lethargy. These symptoms are temporarily related to working in or occupying a particular building. Although these symptoms improve within couple of hours after leaving the building, but this could decrease job satisfaction, work productivity and cause conomic losses (Seppanen and Fisk, 2006).

House Plants for Relieving Sick Building Syndrome

To improve their living and work spaces, people plants indoors as plants serve as the symbol of nature and bringing nature indoor is the best possible way to get rid of this problem. Although, technology advancements for improving indoor air quality are going on (e.g., electronic filtering systems); however, these are often expensive and not widely used. So, biological methods, such as use of plants are increasingly being used. Plants are known to absorb air pollutants via their stomata during normal gas exchange. Several pollutants have been shown to be sequestered or degraded in situ or after transfer to other locations having plants (Schmitz *et al.*, 2000). Some major air pollutants are also removed by absorption or adsorption to the plant surface, microorganisms or soil particles (Orwell *et al.*, 2004). In addition, indoor plants consume little energy, absorb carbon dioxide, decelerate global warming, and contribute to ecological diversity. Thus, they contribute to both public health and sustainable development. Planting beautiful and aesthetic looking indoor plants not only has a picturesque effect but also purifies the surrounding air and give a relaxing feeling. Awaking fresh not only sparkles the morning but also makes the whole day good.

National Aeronautics and Space Administration (NASA) has conducted research on various plants that will better sleep by removing the intoxicants like formaldehyde, benzene, neutralizing the moulds and microbes, increasing the content of oxygen and providing sweet smell which has soothing effect on mind to induce sleep. Several indoor plant species for example, *Spathiphllum willisii*, *Dracaena deremensis* and *Kalanchoe blossfeldiana* can effectively remove benzene while *Chrysalidocarpus lutescens* and *Phoenix roebelenii* can readily remove toluene and xylene. Instead of using sweet smelling chemicals in homes, it is far much better to go natural.

Benefits of Indoor Plants
1. Energy healing potency
2. Provides healthy indoor environment
3. Psychological benefits
4. Prevent allergies by removing allergens
5. Spiritual aspects
6. Increases work efficiency of employees
7. Beneficial effects on children
8. Helps in improving health of patients
9. Absorbs noise
10. Increases humidity of surrounding air
11. Cooling effect
12. Aesthetic and economic values

Plants for a better Sleep :

NASA (National Aeronautics and Space Administration) has recommended various plants to decorate bedroom to have a sound sleep (http://youtu.be). Some of these are:

- Jasmine: It is having gentle and soothing effect on mind.
- Lavender (*Lavandula*): The scent of lavender slows down heart rate, reduces blood pressure and lowers down mental stress. It also reduces crying in babies and put them to deep sleep.
- Snake plant (*Sansevieria trifasciata*): It improves air quality by filtering certain toxins like formaldehyde, benzene. It also takes out CO_2 and increases O_2 content and requires no extra care.
- *Aloe vera*: It gives O_2 during night. NASA reported that it is the best plant for improving air quality. It requires more sunlight, so should be placed near window.
- Gardenia: It reduces stress but a little bit difficult to manage. It should not be placed in direct sunlight but on bright light.
- Spider plant (*Chlorophytum comosum*): NASA reported that spider plant cleans 90% of carcinogenic formaldehyde fumes and bad odours.
- English ivy (*Hedera helix*): It reduces allergy and asthma and also reduces 78% of air borne moulds within 12 hours. It requires moderate sunlight and should be placed in hanging baskets as its leaves are poisonous to children and pets.
- Peace lily (*Spathiphyllum wallisii*): It removes intoxicants and increases humidity by 5%. NASA called this plant as 'superstar'. It also neutralizes air borne microbes.
- Gerbera daisies: It has beautiful and colourful flowers and prevents allergy.
- Golden pathos (*Epipremnum aureum*): It is also known to remove intoxicants but is poisonous so should be placed out of reach of children and pets.

Conclusion

As people spend most of their time indoors, they are faced with increasingly severe physical and mental health and wellbeing problems due to poor indoor air quality. The quality of indoor air is degraded by the emission of several indoor air pollutants which have a detrimental effect on mental fitness, job satisfaction, work productivity and overall well-being. Sick building syndrome is a serious health issue in this modern urbanized world affecting the people confined in their four walls of the house. This is not only a matter of public health but also of economic, societal, and environmental sustainability. The most effective way for solving this problem is to come close to the nature by bringing nature indoor. The use of indoor plants for removing the dearth of greenery deserves attention. House plants not only purifies indoor air by degrading indoor pollutants but also helps in relieving sick building syndrome, lowers stress and tension and creates a healthy, peaceful and happy ambience leading to enriched overall well-being.

References
1. Lim, Y. W., Yang, J. Y., Kim, H. H., Lee, Y. G., Kim, Y. S., Jang, S. K., Son, J. R., Roh, Y. M., and Shin, D. C., 2006, "Health risk assessment in terms of VOC at newly built apartment house," J. Kor. Soc. Indoor Environ., 3, pp. 211–223.
2. Yang, X., J. Srebric, X. Li. and G. He. 2004. Performance of three air distribution systems in VOC removal from an area source. Building Environ. 39: 1289-1299.
3. Craighead, J. E., 1995, "Indoor air quality and pollution," In: Craighead, J. E. (ed.). Pathology of Environmental and Occupational Disease. Mosby Year Book, St. Louis, pp. 29–39.
4. Sullivan Jr., J. B., Van Ert, M. D., Krieger, G. R., and Brooks, B. O., 2001, "Indoor environmental quality and health. In: J. B. Sullivan Jr. and G. R. Krieger (eds.). Clinical environmental health and toxic exposures (2nd ed., p)," Lippincott Williams & Wilkins, a Walter Kluwer Co., Philadelphia, PA, USA, pp. 669–704.
5. Wolkoff, P., 2003, "Trends in Europe to reduce the indoor air pollution of VOCs," Indoor Air 13 (6), pp. 5–11.

6. World Health Organization. ((2009Global health risks: mortality and burden of disease attributable to selected major risks. World Health Organization.
7. Shinohara, N., Mizukoshi, A., and Yangisawa, Y, 2004, "Identification of responsible volatile chemicals that induce hypersensitive reactions to multiple chemical sensitivity patients," J. Expo. Anal. Environ. Epidemiol, 14, pp. 84-91.
8. WHO (1983), "Indoor Pollutants: Exposure and Health Effects," Report of a WHO Meeting, World Health Organization Regional Office for Europe, Copenhagen.
9. Seppanen, O., and Fisk, W.J. (2006), "A procedure to estimate the cost-effectiveness of the indoor environment improvements in office work", in Clements-Croome, D. (Ed.), Creating the Productive Workplace, Taylor & Francis, London.
10. Schmitz, H.,Hilgers, U., and Weidner, M., 2000,"Assimilation and metabolism of formaldehyde by leaves appear unlikely to be of value for indoor air purification," New Phytol, 147(2), pp. 307–315.
11. Orwell, R. L., Wood, R. A., Burchett, M. D., Tarran, J., and Torpy, F., 2006, "The Potted-Plant Microcosm Substantially Reduces Indoor Air VOC Pollution: II. Laboratory Study, Water, Air, Soil Pollution," 177, pp. 59–80.
12. http://youtu.be/2PNG7iFgf2c/9 Sep 2019.

Assistant Professor
Govt. Girls College, Sadulshahar (Sri Ganganagar)
emai : madhu_parmar73@rediffmail.com

18. Duration of Parental Care in Blackbucks in Captivity

Sonia Yadav[1] and Dr. Raksha Modi[2]

Abstract

Blackbuck (*Antilope cervicapra*) is an antelope belonging to family Bovidae. It is indigenously found in the plains of India. The behavioural attributes observed in blackbucks are territoriality, lekking, parental care, foraging, etc. The research was conducted to study the detailed analysis of the duration of parental care in blackbucks in the captivity. The study area selected is Kamla Nehru Prani Sangrahalaya, Indore, Madhya Pradesh. The present investigation was done using visual observations. Parental care is a type of behavioural and evolutionary strategy adopted by animals. It involves parental investment made to the evolutionary fitness of offspring. Parental care differs in different animal groups in terms of how parents care for offspring, and the amount of resources invested by parents. The parental care is generally displayed in blackbucks till 2 months of age. The blackbuck juveniles upto the age of about 2 months remain closely connected to their mother. The type of parental care seen in blackbucks is maternal care as only mothers care for the young and the fathers (male parents) have no role in parental care. The first fifteen days of the fawns are wholly spent with the mother blackbuck and in the next fifteen days fawns were seen performing various activities with their mother which include following, playing, feeding, etc. The duration of parental care and beginning of weaning was observed in 10 subject fawns and the average was calculated. The duration of parental care in blackbucks is about 2 months in captivity. Soon after weaning begins, the parental care ceases.

Keywords : Captivity, behavioural, weaning, maternal, duration

Introduction

Blackbuck (*Antilope cervicapra*) is an antelope belonging to family Bovidae. It is indigenously found in the plains of India. Blackbucks are completely Indian sub-continental in terms of distribution

(Rahmani, 1991). The blackbuck belongs to the same tribe (Antilopini) that includes the springbok, the gerenuk and gazelles. Wild Blackbuck population is likely to continue to decline; the population of blackbucks is declining as a consequence of poaching, hunting, shrinkage of natural habitats and deforestation, and predation with probably fewer than 25,000 individuals in their native range. The Blackbuck acts as an indicator for natural disturbances in the forest habitat (Rajagopal, 2009). Adult male blackbucks have long horns that are V-shaped and spirally twisted. There is a wide difference seen in coloration of adult males and females. Adult males are black and white in color while the adult females are reddish yellow in color. Immature males also look reddish yellow in color. This type of contrast in coloration is not seen in any of the blackbuck's tribal relatives. Blackbuck is also known as the Indian antelope. It is very commonly found in India, Nepal, and Pakistan.

Blackbucks are primarily grazers and form herds which are characteristically loose and unstable associations; the number of individuals in herds can range from less than ten individuals to several hundred (Rajagopal and Archunan, 2016). Prasad (1985) stated that blackbucks spend a large amount of time in foraging almost 34%. Chiefly blackbucks are found frequently in open short grassy regions. Generally, woodlands and shrublands are avoided by them. When the grass is sparse, they browse for it as they love green grass. In the semidesert areas of Rajasthan, the blackbucks drink water twice a day. Day time is the time when they are most active and seek shade for hardly 2 to 3 hrs in mid-day. They can normally tolerate hottest sun. The blackbucks are called diurnal animals. Meena et al. (2017) reported that blackbucks are mainly active during the day; they live in small herds of around twenty to thirty members most of the time.

Major behavioural activities observed in blackbucks are foraging, walking, standing and resting; foraging includes both browsing and grazing. Three types of groups, generally small are found. They are of females, males and bachelor herds. Males attract females for mating by lekking. Lek is a gathering of male individuals that engage in competitive displays and rituals of courtship to entice females for mating. Other males are restricted from such places. Lekking takes place at spots where females go for foraging. So, at those places males can attempt to get a mate and can undergo mating. Blackbucks are herbivores and mostly prefer low grasses and found browsing for substantial time.

Blackbucks have prominent eyesight by which they avoid capture. Blackbucks run very fast, so that they outrun most of their predators. Blackbuck is the major wild prey of wolf in India. The only predator which blackbucks can't outrun is the cheetah. Mughals once used cheetah for the sport of coursing blackbucks and gazelles. The chief predators of blackbuck now are jackals and pariah dogs. The captive Indian Blackbuck also varies highly within captive populations; there is a need to increase their breeding in captivity. Extremely high levels of visitor density influenced various behaviours; the affected behaviours were moving, resting, reproductive, social, and aggressive behaviors (Rajagopal et al., 2011). The behaviour of an animal is influenced by environmental factors as well as social interactions (Isvaran, 2005).

Study area selected is Kamla Nehru Prani Sangrahalaya, Indore, Madhya Pradesh. Indore, formerly spelled Indur, is a city located in western part of Madhya Pradesh state, central India. Indore is the most populous city and the largest city of Madhya Pradesh; it serves as the headquarters of both Indore District and Indore Division. Indore is situated at an average altitude of 553 meters (1,814 ft) above sea level and located on the southern edge of Malwa Plateau. Among the major cities of Central India, it has the highest elevation. Indore is situated 190 km (120 mi) west to Bhopal; Bhopal is the state capital of Madhya Pradesh. According to 2011 census, estimated population of Indore is 1,994,397 (municipal corporation) and 3,570,295 (urban agglomeration). The land area of the city is 530 square kilometres (200 sq mi); this value makes Indore the most densely populated major city in the central province. Indore zoo or Kamla Nehru Prani Sangrahalaya is a zoological park situated in Navlakha, Indore, Madhya Pradesh; it is managed completely by Indore Municipal Corporation which looks after its management, maintenance and administration. Kamla Nehru Prani Sangrahalaya is a well recognised zoo, even recognised by CZA (Central zoo authority of India). Indore zoo is also known as "Land to many wildlife wonders". Indore zoo is the most advanced zoo of India for many reasons; it deploys online booking of tickets and it powers animal health app. The blackbucks are also introduced to the Ralamandal Wildlife Sanctuary. The Ralamandal Wildlife Sanctuary is located in Indore, Madhya Pradesh; Madhya pradesh is the state located in the heart of India. The Ralamandal Wildlife Sanctuary was established in 1989 and one of the most popular wildlife reserves in India.

Materials and Methods

The research was conducted to study the detailed analysis of the duration of parental care in blackbucks (Antilope cervicapra) in the captivity (Indore zoo). The present study was mainly based on the behavioural attributes of blackbuck relating to parental care; study was mainly conducted to find the duration of the parental care provided to the fawns. The present investigation was done using visual observations. The field binoculars were the important tools used throughout the study to conduct observations. All the important activities were keenly observed and noted down. In this study, certain activity refers to any action which resulted into a change in the position of the creature in relation to space. Whenever a group or individual animal came into the sight the observational information were recorded in the data sheet. A safe distance was maintained for observing the animals. The changes in the various aspects of behaviour were also studied taking various parameters in consultation such as activities which are performed exclusively by fawns, the companionship of fawns with mother and many other activities. Cameras were used at different spots and different angles and their activities were monitored. For live observation, binoculars were used. With the help of binoculars, the objects located far from us can also be observed clearly.

Results and Discussions

Parental care is a type of behavioural and evolutionary strategy adopted by animals. It involves parental investment made to the evolutionary fitness of offspring. Patterns of parental care are widespread and highly diverse across the animal kingdom (Kokko and Jennions, 2008). Parental care differs in different animal groups in terms of how parents care for offspring, and the amount of resources invested by parents. For illustration, there may be significant variation in the amount of care invested by each parent. In some animals, mother invest more in parental care while in some fathers care more for the offsprings and in some investment may be shared equally by both the parents. Gonzalez-Voyer and Kolm (2010) reported that many hypotheses have been proposed to describe this variation and patterns in parental care that exist between the sexes, as well as among species. Parental care is beneficial in cases when it increases the parent's inclusive fitness by activities such as by improving offspring survival, quality, or reproductive success (Klug and Bonsall, 2010). Parents make sure that any investment is well-spent because parental care is costly and often affects the parent's own future survival and reproductive success. Parental care is the set of behaviours that contributes to offspring survival. It includes activities such as building a nest, provisioning offspring with food, or defending offspring from predators. Young of reptiles need almost no care as they are produced self-sufficient but some hatchling birds may be helpless at birth, relying on their parents for survival. Parental care may be of mainly four types: maternal or paternal care, biparental care and alloparental care (Kokko and Jennions, 2008). Paternal care is the type of parental care where the main investment is done by the male parent for the offsprings fitness and health; it is most commonly observed in very few bird species where males care for the eggs after they are laid. Maternal care is the type of parental care where mothers care for the offsprings the most; the most evident form of maternal care is seen in mammals. In mammals, the female parent (mother) undergoes gestation and also shows lactation; gestation is the period from fertilization of gametes to form zygote to delivery of the young and lactation is feeding the mother's milk to the young as preliminary food. Biparental care is the type of parental care in which both parents (mother and father) contribute almost equally for the care of the young. Remes et al. (2015) noted that this type of parental care is common in animals where the adult sex ratio is quite balanced.

Silver (1983) reported that nearly all the species of birds have monogamous relationships and they display biparental care. Alloparenting is a type of parental care in which the parents care for the youngs that are not theirs; the offsprings are non-descendants of the parents. This type of care is selfless but it costs the parents the reproductive benefits. Mating may cause sexual conflict and further familial conflicts may continue after mating when there is parental care of the eggs or young. Conflicts may arise between male and female parents over how much care each should provide; conflict may arise between siblings also over how much care each should demand, and conflicts may arise between parents and offspring over the supply and demand of care (Parker et al., 2002). The evolutionary fitness of the offspring receiving the care is benefitted, but it produces a cost for the parent organism as parents spend energy on caring for the offspring; this may result into loss of mating opportunities (Bednekoff, 2010; Fox et al., 2018). Parental care is a costly mechanism so it only evolves only when the costs are outweighed by the benefits (Klug and Bonsall, 2014).

Blackbucks are the mammals that belong to family Bovidae; they inhabit grassy plains and lightly forested areas. Blackbucks show clear sexual dimorphism as the males have long ringed horns and their body is dark brown coloured with white colour in underparts and the insides of the legs while the females and fawns are very light brown colored. The blackbucks are usually sexually active throughout the year and the litter size is 1. The female blackbucks give birth to usually one young at a time.

The parental care in blackbucks was well observed; the parental care is generally displayed in blackbucks till 2 months of age. The blackbuck juveniles upto the age of about 2 months remain closely connected to their mother. The type of parental care seen in blackbucks is maternal care as only mothers care for the young and the fathers (male parents) have no role in parental care. The relationships are not monogamous but polygamous in blackbucks. The male blackbuck distances itself from the female blackbuck just after the mating. The mother takes care of the offspring alone for around 2 months. For the time period of those 2 months, the fawns completely derive their nutrition from mother's milk. Lactation is the process of secretion and yielding of milk by females after giving birth. The milk produced during lactation is contained in the mammary glands; the growth of mammary glands takes place during gestation due to the ovarian and placental hormones (Donovan, 2022). The milk is produced in the mammary glands during gestation but is held in them without release, by the action of combination of estrogen and progesterone; estrogen and progesterone inhibit milk secretion by blocking the release of prolactin from the pituitary gland and by making the mammary gland cells unresponsive to this pituitary hormone (Donovan, 2022). The stimulus of nursing or suckling supports constant lactation; it is done by the secretion of prolactin and also leads to release of oxytocin, which causes the contraction of muscles around the alveoli in the breast and ensures the ejection of milk (Donovan, 2022). After about 2 months, weaning starts in fawns. Weaning is the act of switching from mother's milk to other food options by the fawn. Once the weaning begins and the fawn gets accustomed to the outside food options, the mother moves away from the fawn. This marks the end of the parental care in blackbucks.

The fawns, till about 15 days of the age, spend their whole time in company of their mother and it is hard to find such young fawns away from their mothers. In zoo, the environment is free from predators, so there is no chance of spotting any young fawn without his mother; however, in wild, the young fawns may be found alone or without their mother in case when their mother is being preyed by some predators. The male blackbuck moves away from the female just after the copulation, so the males never get connected to neither the females nor their offsprings. This is because of the existence of polygamous relationships in blackbucks. The first fifteen days of the fawns are wholly spent with the mother blackbuck and after this period, the fawns develop a little more strength so, they were observed performing various activities with their mother which include following, playing, feeding, etc. The duration of parental care and beginning of weaning was observed in 10 subject fawns. The values were recorded in a table and average was calculated.

Duration of parental care received by 10 subject fawns	
Subject Fawns	Duration of parental care (in days)
Fawn 1	57
Fawn 2	72
Fawn 3	65
Fawn 4	60
Fawn 5	58
Fawn 6	66
Fawn 7	69
Fawn 8	62
Fawn 9	70
Fawn 10	59

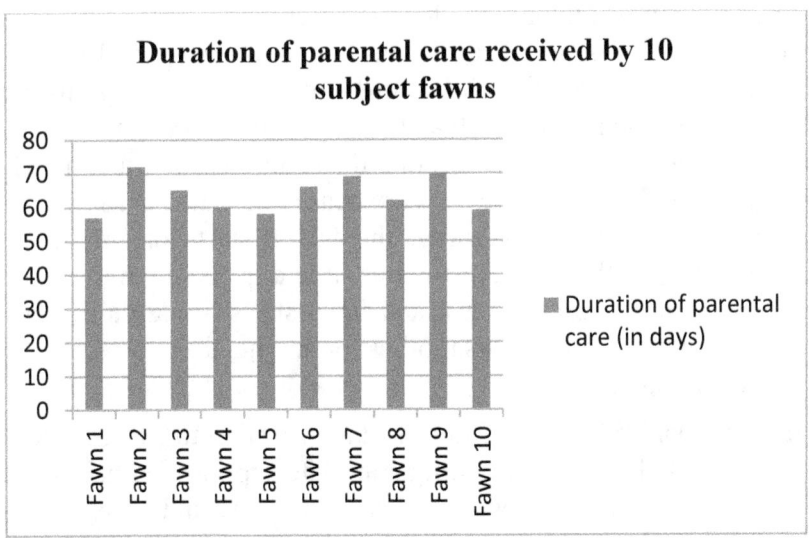

The average duration of parental care in blackbucks in zoo was 63.6 days which is nearly equal to 2 months. There is a difference in days of parental care received due to early weaning in some fawns (such as in Fawn 1) and delayed weaning in other fawns (such as in Fawn 2). Though there is a good difference in cessation of lactation and weaning in fawns, the physical health of the fawns is not affected by it. At zoos, the animals are well-nourished so, the fitness of a fawn which receives parental care for few days more is almost same as the fitness of a fawn which receives parental care for few days less than others. After the beginning of weaning, the parental care faded and then disappeared as the female (mother of the fawn) left the fawn and resumed its other behaviours except the parental care.

Conclusion

The blackbuck (*Antilope cervicapra*) are mammals that belong to family Bovidae and are exclusively found in the Indian subcontinent. The blackbucks show a clear sexual dimorphism as the males have horns while usually females lack horns and also there is huge difference in colouration in males and females. The blackbucks mate round the year and they give birth to just one offspring at a time. The offspring is being cared by its mother only as the blackbucks show maternal care. Maternal care is a type of parental care in which only the female parent takes care of the young to support its fitness and health. The young, during first 15 days of its birth, remains stick to its mother; it starts showing different activities with the mother such as feeding, playing, etc., after the age of 15 days. For the first about 2 months of its life, the fawn is taken care by its mother. The parental care ceases in blackbucks when the weaning starts. Due to weaning, the fawns start consuming food other than the mother's milk; it is this time during which the mother distances herself from the fawn and fawn starts living its life on its own without depending on its mother for any need.

Acknowledgement

I am very thankful to Dr. Uttam Yadav sir (Director, Kamla Nehru Prani Sangrahalaya) for providing complete support and help in data collection for the purpose of research. I would also like to acknowledge zoo volunteers for the assistance provided in carrying out the research. My heartfelt thanks to my research supervisor Dr. Raksha Modi ma'am (Associate Professor, Department of Zoology, P.M.B. Gujarati science college, Indore) for always guiding and helping me in every step of my research.

References

Bednekoff, P. A. (2010). Life Histories and Predation Risk. Encyclopedia of Animal Behavior. Academic Press. pp. 285-286.

Donovan, B. T. (2022). Lactation. Encyclopedia Britannica. https://www.britannica.com/science/lactation

Fox, R. J., Head, M. L. and Barber, I. (2018). Good Parenting May Not Increase Reproductive Success Under Environmental Extremes. Journal of Evolutionary Biology. doi: 10.1111/jeb.13358.

Gonzalez-Voyer, A. and Kolm, N. (2010). Parental Care and Investment. Encyclopedia of Life Sciences. doi: 10.1002/9780470015902.a0021907.

Isvaran, K., (2005). Variation in male mating behaviour within ungulate populations: patterns and processes. Current Science, 89(7), pp.1192–1199. doi: https://www.jstor.org/stable/24110971

Klug, H. and Bonsall, M. (2010). Life history and the evolution of parental care. Evolution, 64(3), pp.823-835. doi: 10.1111/j.1558-5646.2009.00854.x.

Klug, H. and Bonsall, M. B. (2014). What are the benefits of parental care? The importance of parental effects on developmental rate. Ecology & Evolution. 4 (12): 2330–2351. doi: 10.1002/ece3.1083.

Kokko, H. and Jennions, M.D. (2008). Parental investment, sexual selection and sex ratios. Journal of Evolutionary Biology, 21, pp.919–948. doi: 10.1111/j.1420-9101.2008.01540.x.

Meena, R., Saran, R.P., Chourasia, V. (2017). Population Characteristics, Habitat Availability, Forage Preferences and Threats to the Blackbuck *Antilope cervicapra* (Linn) in the Sorsan Region of Baran, Rajasthan. World Journal of Zoology, 12, pp. 53-59. doi: 10.5829/idosi.wjz.2017.53.59

Parker, G.A., Royle, N.J. and Hartley, I.R. (2002) Intrafamilial conflict and parental investment: a synthesis. Philosophical Transactions of the Royal Society of London. Series B, 357, pp.295–307. doi: 10.1098/rstb.2001.0950.

Prasad, N.L.N.S. (1985). Activity-time budget in blackbuck. Proc Ani Sci, 94, Pp.57–65. doi: 10.1007/BF03186327

Rahmani, A. R, (1991). Present Distribution of the Blackbuck (*Antilope cervicapra*) in India, With Special Emphasis on the Lesser Known Populations. Journal of Bombay Natural History Society, 73. 35-45. doi: https://biostor.org/reference/152949

Rajagopal, T. (2009). A study on the reproductive behaviour and pheromones of an endangered Indian Blackbuck (*Antilope cervicapra* L.) to enhance captive breeding and conservation. Ph.D. thesis, Bharathidasan University, Tiruchirappalli, India. doi: 10.12944/CWE.4.1.18

Rajagopal. T., Manimozhi. A., Archunan. G. (2011). Diurnal variation in preorbital gland scent marking behaviour of captive male Indian Blackbuck (*Antilope cervicapra* L.) and its territorial significance. Biol Rhythm Res 42:27–38. doi: 10.1080/09291011003693161

Rajagopal, T., Archunan, G. (2016). Dominance Hierarchy in Indian Blackbuck (Antelope cervicapra L.): Sources, Behavior and Role of Pheromone Signals. In: Schulte, B., Goodwin, T., Ferkin, M. (eds) Chemical Signals in Vertebrates 13, pp.217-228. doi: 10.1007/978-3-319-22026-0_16

Remeš, V., Freckleton, R., Tökölyi, J., Liker, A. and Székely, T. (2015). The evolution of parental cooperation in birds. Proceedings of the National Academy of Sciences, 112(44), pp.13603-13608. doi: 10.1073/pnas.1512599112.

Silver, R. (1983). Symbiosis in Parent-Offspring Interactions. Boston, MA: Springer US, pp.145-171. doi: 10.1007/978-1-4684-4565-7_7.

Vats, R. & Bhardwaj, C.S. (2009). "A study of reproductive behaviour of Indian black buck (Antilope cervicapra) Linn. With reference to courtship, breeding, fawning and colouration" (PDF). Current World Environment. 4 (1): 121–125. doi:10.12944/CWE.4.1.18.

[1]**Ph.D. scholar,**
DAVV University, Indore,
Madhya Pradesh, India
soniaishan1996@gmail.com
[2]**Associate professor,**
Department of Zoology,
P.M.B. Gujarati science college, Indore,
Madhya Pradesh, India
email : c.rakshaa@gmail.com

19. Impact and use of Fertilizers in Indian Agriculture

Ms. Komal Bansal

After the independence the major challenge in front of India is to produce sufficient food for growing population. Our first five year plan also focused on agriculture. For attaining the target of food grain production government focused on green revolution. With the Green revolution, there has been an increase in the consumption of fertilizers in agriculture. Fertilizer is an important input for agriculture production in India. Fertilizers are substances containing chemical elements such as manure a mixture of nitrates that improves the growth of plants. They give nutrition to the crops. Fertilizers are used for increasing yields and growing plants. India ranks second in the world and first among the South Asian Association of Regional Cooperation countries in terms of total fertilizer consumption.

The paper throws light upon the impact of the overuse or imbalanced application of fertilizer nutrients on crop productivity. The paper highlights the advantages and disadvantages of fertilizers and role of fertilizers in agriculture sector. The present paper emphasizes the government initiatives and government policies regarding fertilizers. The paper concludes with that sustainability in agriculture can be maintained through rational use of fertilizer use and organic farming.

Keywords : Organic Farming, Sustainable Agriculture, Environment Degradation, Fertilizers

Introduction :

"The secret of rapid agricultural progress in the under developed countries is to be found much more in agricultural extension, in fertilizers, in new seeds, in pesticides and in water supplies than in altering the size of the farm, in introducing machinery, or in getting rid of middle men in the marketing process". W.A.Lewis

The benefits of agriculture have been immense. Before the dawn of agriculture, the hunter gatherer lifestyle supported about 4 million people globally. Modern agriculture now feeds 6,000 million people. Global cereal production has doubled in the past 40 years, mainly from the increased yields resulting from greater inputs of fertilizer, water and pesticides, new crop strains, and other technologies of the 'Green Revolution'. This has increased the global per capita food supply, reducing hunger, improving nutrition and sparing natural ecosystems from conversion to agriculture. By 2050, global population is projected to be 50% larger than at present and global grain demand is projected to double. This doubling will result from a projected 2.4-fold increase in per capita real income and from dietary shifts towards a higher proportion of meat associated with higher income. Further increases in agricultural output are essential for global political and social stability and equity. Doubling food production again, and sustaining food production at this level, is major challenges. Doing so in ways that do not compromise environmental integrity and public health is a greater challenge still. There are scientific and policy challenges that must be met to sustain and increase the net societal benefits of intensive agricultural production.

Why do we need Fertilizers?

The increasing number of population requires ample amount of food. The main obstacle in higher food grain production is pests, loss of soil fertility and lack of nutrients. To sustain the needs of people there should be proper work to vanish these problems. Hence the importance of using fertilizer rises. Fertilizers are used to enhance the growth of plants. Fertilizers are substances containing chemical elements such as manure a mixture of nitrates that improves the growth of plants. They give nutrition to the crops. Fertilizers are used for increasing yields and growing plants. They incorporate mainly three nutrients. Nitrogen acts as growth booster which can be characterized by the green color of plants. Phosphorus substance in fertilizer aids is the faster formation of seed and root development. Potassium nutrient are used for strong stem growth, movement of water in plants, promotion of flowering and fruiting. Three secondary micronutrients are Calcium, Magnesium, Sulphur.

Advantages and Disadvantages of Using Chemical Fertilizers :
Advantages

Fertilizer contributes to 50% crop production in India. Proper use of fertilizer enables to solve the problem of starvation. It is the source of from which plants grow and most plant nutrients. The first advantage of fertilizer is that balanced use of fertilizer improves soil health which will increase crop productivity and augments farmer profit. Plants needs almost 17 nutrients from which 3 obtained from water and soil and others from soil. Crop's responsiveness to fertilizer is maximised and environmental impact of fertilizer is reduced when crops are managed for improved nutrient efficiency through Best Management Practices (BMPs) which balance production inputs at the appropriate levels.

Other advantages of fertilizers are :

1. Nutrients are soluble and immediately available to the plants; therefore the effect is usually direct and fast.
2. The price is lower and more competitive than organic fertilizer, which makes it more acceptable and often applied by users.
3. They are quite high in nutrient content; only relatively small amounts are required for crop growth.

Disadvantages

1. Over application can result in negative effects such as leaching, pollution of water resources, destruction of microorganisms and friendly insects, crop susceptibility to disease attack, acidification or alkalization of the soil or reduction in soil fertility thus causing irreparable dam age to the overall system.
2. Oversupply of N leads to softening of plant tissue resulting in plants that are more sensitive to diseases and pests.
3. They reduce the colonization of plant roots with mycorrhizae and inhibit symbiotic N fixation by rhizobia due to high N fertilization.
4. They enhance the decomposition of soil OM, which leads to degradation of soil structure.

5. Nutrients are easily lost from soils through fixation, leaching or gas emission and can lead to reduced fertilizer efficiency.

Role of Fertilizers in Agricultural Sector :

Agriculture productivity is dependent upon various factors like soil properties, climate conditions, irrigation facilities, seed quality ad variety, cropping pattern, techniques of farming, prevention from pests but more importantly usage of optimum primary, secondary and micronutrients.Thus, the role of Government become more significant in making available all types of nutrient at affordable to farmers.

Chemical fertilizers have played an important role in making the country self reliant in food grain production. The role of Government of India has been significant because government has been consistently pursuing policies conductive to increased availability and consumption of fertilizers at affordable prices in the country. This is the reason the annual consumption of fertilizers in nutrient terms (N, P&K), Has increased from .07 million MT in 1951-52 to more than 28 million MT in 2010-11 and per hectare cconsumption, has increased from less than 1 Kg in 1951-52 to the level of 135 Kg now.

There is no denying the fact that over the years increased usage of fertilizer increases the agriculture productivity. Current trends in agricultural output, however, depict that the marginal productivity of soil in relation to the application of fertilizer is declining. The comparatively high usage of straight fertilizers (Urea, DAP & MOP) as against the complex fertilizers (NPK) which are considered to be agronomically better including low or non usage of secondary and micro nutrients has also probably contributed towards slowdown in growth of productivity. The declining fertilizer use efficiency is also one of the factors for low productivity.

The Status of Consumption and Production of Fertilizers in India

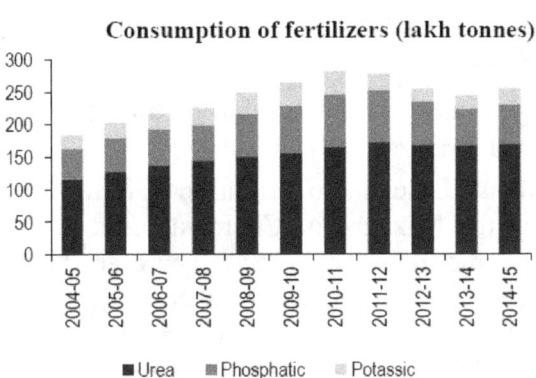

Consumption of fertilizers (lakh tonnes)

■ Urea ■ Phosphatic ■ Potassic

Sources: Agricultural Statistics at a Glance 2015; PRS.

Consumption of Fertilizers from 1999-2010 (thousand tons of nutrients)

Year	Consumption	Change in the percentage of consumption
1999-2000	180.69	
2000-2001	167.02	-7.56
2001-2002	173.59	3.93
2002-2003	160.94	-7.28
2003-2004	167.99	4.38
2004-2005	183.99	9.52
2005-2006	203.4	10.54
2006-2007	216.51	6.44
2007-2008	225.70	4.24
2008-2009	249.09	10.36
2009-2010	264.86	6.33

Source : Economic Survey 2010-11

The above table shows that fertilizer consumption from the year 1999-2000 to 2009-10. There was only two years which shows negative growth rate in fertilizer consumption i.e 2000-01, 2002-03. Another year shows increasing trend of fertilizer consumption in India. There was a very big increase in fertilizer consumption in the year 2005-06, the fertilizer consumption was 203.4 thousand tonnes of nutrients in India. 10.54% growth rate recorded during this year.

NPK Fertilizer :

NPK rating explains the amount of nitrogen, phosphorus, and potassium in a fertilizer.

The manufacture, sale, and distribution of fertilizers in the country is regulated by the Ministry of Chemicals and Fertilizers, under the Essential Commodities Act, 1955. There are three major types of nutrients used as fertilizers: Nitrogen (N), Phosphatic (P), and Potassic (K). Of these, the pricing of urea is controlled by the government, while P and K fertilizers were decontrolled in 1992, on the recommendation of a Joint Parliamentary Committee. It has been observed that urea is used more than other fertilizers. While the recommended ratio of use of the NPK fertilizers is 4:2:1, this ratio in India is currently at 6.7:2.4:1.6 Overuse of urea is especially observed in the states of Punjab, Haryana and Uttar Pradesh. An imbalanced use of urea may lead to a loss of fertility in the soil over a period of time, affecting productivity. Urea (N) is the most produced (86%), consumed (74%) and imported (52%) fertilizer in the country. The government determines the quantity of fertilizers to be imported based on their domestic availability.[i]

Fertilizer Subsidy :

The cost of the fertilizer is very high in many countries. Farmers can't afford the price of fertilizers so they will not be able to apply nutrients in their fields. The result was lower crop production and lower food grains. In many developing countries government took step in the form of subsidy to reduce the price of fertilizers. The main objective of the government is to promote agricultural development through wider adoption and economically efficient use of fertilizers. There has been a growing trend of subsidy burden on the Government. Higher amount of subsidy given by government to increase in consumption, increase production, higher input cost, increasing import prices of fertilizers as well as feedstock and intermediaries and more importantly keeping the retail price of the f ertilizers at affordable levels. To promote the use of fertilizers by farmers, the central government provides a fertilizer subsidy to the producers of fertilizers.

Improper use of Fertilizers and Decline in Crop Productivity :

Fertilizers should be properly used in agriculture and application of fertilizers based on scientific testing of soil. Only scientific fertilizers should be used in field to augment missing or deficient nutrient in soil. The inadequate or imbalance nutrient use coupled with neglect of organic manures has caused multi nutrient deficiencies in many areas.

The Impact of Fertilizer on Environment :

Inadequate use of fertilizer can cause harmful effect on environment. If fertilizers are used with proper scientific soil test then it would enhance agricultural productivity. Indiscriminate use of fertilizers and unawareness of farmers have led to several problems regarding environment and soil health.

Fertilizer Policy

There is increasing the trend of fertilizer consumption in agriculture. This gives benefit to fertilizer industry also. The fertilizers consumption is affected by various factors like irrigation, high yielding variety seeds, size of the farm credit etc. Government provides fertilizer subsidy to provide fertilizers to farmers at remunerative prices. Soil health card schemes and various policies introduced by the government to aware farmer about proper use of fertilizers.

Since Independence government of India has been regulated sale, price and quality of fertilizers. Government of India has passed Fertilizer control order under Essential commodity act in the year 1957.There was no provision of subsidy till 1977. Only potash for which subsidy was paid only for a year in 1977. Maratha committee was constituted to recommend on fertilizer subsidy. Government accepted the recommendations and introduced retention price scheme for nitrogenous fertilizer in 1977. In 1990 when India was facing huge fiscal deficit and there was an impending danger of foreign exchange crisis, Government announced an increase in 40% in the price of fertilizers in july 1991. Some of the fertilizers which were under the subsidy scheme were decontrolled. But in the context of present scenario means for sustained agricultural growth and to promote balanced nutrient growth, it is necessary to provide farmers fertilizers at affordable rate. Urea is the only controlled fertilizer which is sold statutory notified uniform sale price, and decontrolled Phosphate and Potassic Fertilizers are sold at indicative maximum retail prices. Manufacturers of the fertilizers go through the problem of reasonable return on their investment with reference to controlled prices. The solution of this problem is introduced with Fertilizers policies. A scheme which is named as New Pricing scheme for urea units and the concession scheme for decontrolled Phosphatic and Potassic Fertilizers became beneficial for agricultural point of view. NPS-1 for urea was introduced in 1 april 2003. The objective of this policy was to encourage efficiency parameters of international standards based on the usage of the most efficient feedstock.

Economics of Fertilizer use :

The first thing farmers have to decide that what type of fertilizers have to use and how to apply. They should spend money on fertilizers with the thought that how much they will gain. The primary test is an estimation of the crop production and its value that will result from the application of a given quantity of the plant nutrients. Because law of diminishing returns applicable in agriculture sector of India. The aim of the farmer is to use fertilizer at most profitable and economic optimum rate. Thus economics of fertilizer use helps us to analyze the benefits and losses of fertilizer use. A farmer induced to apply Fertilizer if the profits from the fertilizers greater than fertilizer cost. A favorable cost price relationship encourages higher fertilizer usage. An unfavorable relationship restricted greater fertilizer use and result of this seen in the form of lower crop production and lower yields. Reflecting different production cost, import costs, government policies fertilizer prices were vary from country to country.

Conclusion :

Fertilizer along with improved seed use is the key driver to agricultural production which in turn drives the attainment of food security. Fertilizers are critical in improving agricultural production and food security through nutrient loss replenishment on farmer's fields. Food security exists when all people at all times have access to sufficient, safe and nutritious food to maintain active healthy life. Thus the use of agrochemicals is necessary now days. But we should move towards organic farming and adopts techniques that are not harmful for health and increase the production of food grains. India has to put in place a well defined comprehensive system that enforces stringent policies on balanced fertilizer use, besides facilitating integrated nutrient management with locally available organic manures/crop residues and cultivation of efficient crop geno

References :
1. Anderson Henrik et al (2014), "Pesticides and Health: A review of evidence on health effects, valuation of risks related to pesticides, and benefit cost analysis", Tolouse school of economics, France ONLINE.
2. Avneesh Kumar and Kumar Anuj(2016), "Black Face of Green Revolution in Malwa Region of Punjab", *Biological Innovations Research and Developmental Society.*
3. Baishya Karishma(2015), "Impact of Agrochemicals Application on Soil Quality Degradation", *International Journal of Science and Technology,* Vol no 4.
4. Bhardwaj Tulsi and J.P Sharma(2013),"Impact of Pesticides Application in Agricultural Industry: An Indian Scenario", *International journal of agriculture and food science technology,* (Vol 4), 817-822.
5. Bedi Singh Jasbir el al:(2015), "Evalution of Pesticides residues in human blood Samples in Punjab", Online assessed on 21.1.2015
6. Kulshrestha.K.S and Jai Singh Rathore (2016). Agriculture Trends and Development in Rajasthan, Rajasthan Economic Journal , vol38-39(1&2)
7. Tilman David et al;(2002). "Agricultural sustainability and intensive production practices", Nature Publishing Group, Vol 418

8. https:/en Wikipedia.org/wiki
9. Chen Hshuan Jen(2012), "The Combined Use of Chemical and Organic Fertilizers and Biofertilizer For Crop Growth and Soil Fertility."
10. Shukla.K.Arvind(2022), "Fertilizer use in Indian Agriculture and its Impact on human health and environment" Indian Journal of Fertilisers 18 (3) : 218-237
11. Deshpande Tanvi(2017), "State of Agriculture in India"

Assistant Professor, Economics
Dr. Bhimrao Ambedkar Government College,
Sriganganagar
email : Komalbnsl786@gmail.com

20. Sustainable Development : A New Approach towards Mitigation of the Conflict between Development and Environment through the Lens of Indian Judiciary

Dr. Suruchi
Saurabh

Abstract

Sustainable development is not a new concept for the world, it has a long existence in society, but it is the contribution of Brundtland Commission in bringing it as a solution to environmental problems. After the period of 1960s when industrialization and urbanization became the most important component for attaining economic growth and development, it resulted in degradation of the environment and its natural resources and it also became apparent that the development and protection of the environment cannot go hand-in-hand. Both have an inverse relationship but the solution of this conflict has been brought by the Brundtland Commission as sustainable Development which required a balanced and harmonious relationship between these two i.e., environment and Development. This has been also emphasized by the Indian judiciary while dealing with the conflict of environmental problems. This article is an attempt to fill the gap between environment and development through highlighting the crucial role of Indian judiciary in proliferation of the concept of sustainable development.

Introduction

Over the centuries society has been focusing on transforming its developing economy to industrial economy, in order to attain this goal society has moved towards urbanization and industrialization which required exploitation of natural resources. Uses and exploitation of natural resources has been converted into over exploitation of it and further it resulted into long term human sufferings i.e., deforestation, soil erosion, climate change, global warming, ozone depletion and many more.

Development is no doubt, an inevitable part of society but development should go hand in hand with the environment. These two components are two sides of the same coin and cannot be achieved separately. Without concern for the environment, development only becomes a human suffering.

The old notion of the relationship between development and protection of environment was believed to be against each other. It was realized that an environmental degradation in order to achieve economic prosperity is none other than a sign of progress. With the advancement of science and technologies, emerged various environmental issues as a health hazardous problem for human being which could not be ignored. After the Stockholm declaration and Brundtland report it has been realized that economic growth is not the antithesis of the environment but a condition underlying it. The Brundtland report asserts that poverty is the biggest polluter and only economic growth can eliminate poverty by creating capacity to solve environmental problems. Economic and environmental concerns have critical links. The concept of sustainable development has evolved from the linkages between economics and ecology. However, unregulated and unrestricted development activities cannot be permitted as these contradict and negate the element of sustainability. Sustainable development is the development which has environmental concerns. This article mainly focusing in bringing to the fore the gravity of the principle of sustainable development and its substance to mitigate the arising conflict between the developmental goals and environment imperatives. While reveling the significance of sustainable development in today's era, this article highly giving priority to accentuate the role of Indian judiciary in proliferation of sustainable development as a source of mitigation. The article mainly based on the doctrinal methodology. Using this method, the researcher composes a descriptive and detailed analysis of legal rules found in primary sources (cases, statutes, or regulations) along with the secondary sources (books, journals, reports of commissions, etc.) this article mainly organized into three parts: first part is relation to the meaning and definition of sustainable development. This part mainly focuses on discussing and explaining the concept of sustainable development, its components and underlying principles. The second part deals with the leading law cases relating environmental issues. Through the case study researcher made an attempt to array the inter-relationship between these two through the lens of Indian judiciary which exhibit the realization of conservation and preservation of natural resources as a condition precedent to economic growth and development and also insistence the role of Indian judiciary to augmenting the principle of sustainable development in protection of environment. The third part is the conclusion which summed up the definition of development as a real development means development with the consideration of environment not at the cost of it.

Sustainable Development : Meaning & its Components

Sustainable development means "the development that meets the needs of the present without compromising the ability of the future generations to meet their own needs."[1] In other words we can say that it is a new concept of economic growth which requires optimum use of natural resources in such a way that does not harm natural resources and preserve it for a long time so that the coming future is also able to fulfill their own needs with natural resources.

Sustainable development is the only path for conserving and promoting the socio-economic well being of people. The concept of sustainable development rests on the foundation of equity.[2] It is structured in two forms of equity. One is Inter-generational equity and second is intra- generational equity.

Inter-generational equity mandates that the present generation should look at natural resources as a trust which has been passed on to them by ancestors and further it has to be passed on to the future generations for their use. Each generation holds the earth and its natural resources as a trustee which should be passed on from generation to generation.

The theory of Intergenerational equity stipulates that we, the human beings, hold the natural resources in common with all the members of our human species, past, present and future generations. As a member of the present generation we have a duty to take care of natural resources as being a trustee of the natural property and must be passed onto future generations along with the quality of natural resources. It means that the present generation is bound not only to preserve natural resources but also has the corresponding duty to maintain the quality of natural resources. All generations are inherently linked to other generations using the common property of nature. There are three main principles of intergenerational equity:

1. **Conversation of Options :** each generation receives a natural and cultural legacy in trust from previous generations and holds it in trust for future generations. This relationship imposes upon each generation certain planetary obligations to conserve the natural and cultural resource base for future generations and also gives each generation certain planetary rights as beneficiaries of the trust to benefit from the legacy of their ancestors.[3] We require natural resources for solving our problems and satisfying our needs. Similarly, future generations also need these natural resources to fulfill their own needs so it is mandated by this principle to conserve the options for future generation to use natural resources according to their needs.

2. **Conservation of Quality :** "requires the present generation to pass the planet on to future generations "in no worse condition than that in which it was received."[4] It requires the development of predictive indices of environmental quality. A detailed framework must be developed for evaluating net impacts on environmental quality.[5]

3. **Conservation of Access :** " requires that members of the present and future generations have equitable rights of access to the planetary legacies, therefore it is required that the present generation must be conserving this access for future generations. Professor Redgwell explains that the principle of Conservation of Access "reflects a basic trust obligation, namely, the general duty of a trustee to maintain equality between the beneficiaries, and to act impartially between life tenants (the present generation) and 'remaindermen' (future generations)."[6] These principles recognize the right to access and duty to share earth's resources in a constrained action of present generations in doing so (Management of natural resources). While on the other hand, Intra-generational equity signifies equity within and between the countries. It

emphasizes looking upon the problems of sustainability in the light of different economies, environmental, cultural & political circumstances prevailing within and between the countries. The Brundtland report asserts that "the future cannot be common in the sense of being equal, fair and just, when the economic and ecological situations of lower and higher income countries are compared."[7] Intra-generational equity requires that the developed countries should provide funds to developing countries to constitute their capacity to protect the environment. This is mainly based on the realization that we have one planet, two worlds and two economies. The responsibilities of each nation to protect nature are proportionate to their respective economies.

Interpretation of Sustainable Development by Indian Judiciary

The apex court of India recognizes the principle of sustainable development in various cases as a mode of mitigation in the conflict of development and environment. Its importance has been spelled out as a basis for balancing ecological imperatives and developmental goals, in the leading cases of environmental jurisprudence. We will look into the leading case but firstly, let's categories the phases of interpretation in various cases into two periods:

1. **Phase of 1980s :** The *Rural Litigation and Entitlement Kendra. vs. State Of U.P. & Ors, 1985*[8] also known as *Dehradun Mussoorie Hills Quarrying case* was the first of its kind which has issues relating to environment and ecological imbalance which introduced the sharp conflict between development and environment. In this case a letter was written by the group of people to the SC against the excessive and uncontrolled mining in the area of Mussoorie hills which caused hazardous environmental degradation resulting in landslides and blockage of subterranean water channels and springs in the valley. Initially, the court constituted two committees to advise the bench on the technical issues relating to the environment. On the report of the Bhargava Committee, the court ordered the closure of a number of lime-stone quarries. Another committee was set up by the court named as Bandyopadhyay Committee to examine whether a particular

limestone quarry can be allowed to operate in accordance with the relevant statutes, rules and regulations. The Court held in 1988 that all the mines in the area (except for three) ought to remain closed. Mining in the valley was held to have violated the Forest Conservation Act. Economic development could not be done at the cost of environmental degradation. Through this case the apex court emphasised the need of mitigation between the two in the larger interest of society and treated sustainable development as a mode of mitigation.

In *kinkari devi v state of HP 1988 case*[9], Himachal Pradesh HC observed that if industrial growth sought to be achieved by reckless mining resulting in loss of life, loss of property, loss of natural resources and creating ecological imbalance then there may be ultimately no real economic growth and development. This case also involves uncontrolled mining in the hilly areas of HP which was an arbitrary action of state government by indiscriminate granting of leases which resulted into loss of natural resources and degradation of environment. In this case also HC pointed out the importance of sustainable development which can only be achieved with the consideration of the environment.

2. **Phase of 1990s :** during the period of 1990s the Supreme Court became more vocal regarding the importance of sustainable development in the era of industrialization and urbanization. This can be seen in the *Vellore Citizen's Welfare Forum v UOI 1996*[10] also known as *Tamil Nadu Tanneries Case*. In the case a number of Tanneries effluents discharged into agricultural fields, road- sides, water- ways and open land which resulted in the blockage of water supply in the city and also deteriorated potable water. Despite state and central government efforts most of the tanneries hardly take any steps to control the pollution and finally a writ has been filed by the forum and the Supreme Court held that the concept of 'Sustainable Development' was a part of customary international law to strike balance between ecology and development. The Supreme Court further directed the tanneries to set up common effluent treatment plants and also obtains NOC from the state Pollution Control Board and those whom NOC was refused remain closed. Through this case the apex court of India recognizes the principle of sustainable

development as a mean for the mitigation of conflict arising between development and preservation of ecology. Rejecting the old notion that development and environment cannot go together, the Supreme Court gave a landmark judgment and held that sustainable development is a viable concept to eliminate poverty and assist to raise the quality of the environment.

Conclusion

At the beginning of the era of industrialization the nations ran towards economic growth and development. In order to achieve this purpose humans have exploited its natural resources uninterruptedly and in an uncontrolled way which resulted in damage to the environment and brought us to the front of environmental problems. Whenever society moves ahead for development it will automatically be based upon the exploitation of natural resources. It was believed that developmental goals and environment protection have anti relationship and society has to choose whether they would prefer development and growth of the society which ultimately generates employment, economy boom, increase living standards of human being etc., or they would prefer to protect natural resources and environment which is essential for human survival. Today each and every country is facing environmental problems which are growing rapidly day by day and if not controlled in a better way it will lead us to an unforeseen menace or we can say irreparable damage which cannot be repaired or reversed. After discussing relevant facts in relation to the environment and sustainable development especially the opinion of apex courts regarding the conflict between environment imperatives and developmental goals. The report of Brundtland commission which was named as 'Our Common Future' provided a mode of mitigation for the conflict between these two components in the form of Sustainable Development. to sum up the article with the words of one landmark judgment of the hon'ble Supreme court in which it was explicitly expressed that without consideration of environment development would only be mere a myth. No real development could be achieved unless and until protection of environment become condition precedent for such kind of developmental goals. If we protect the environment, developmental goals could be achieved otherwise it could only lead us to the path of human suffering for long period.

References
1. "Our Common Future: Report of the World Commission on ...." https://sustainabledevelopment.un.org/content/documents/5987our-common-future.pdf. Accessed 1 Sep. 2022.
2. "Sustainable Development and Equity - IPCC." https://www.ipcc.ch/site/assets/uploads/2018/02/ipcc_wg3_ar5_chapter4.pdf. Accessed 1 Sep. 2022.
3. "Revisiting the Doctrine of Intergenerational Equity in Global ... - CORE." 4 Jan. 2007 https://core.ac.uk/download/pdf/288305264.pdf. Accessed 28 Sep. 2022.
4. "Our Rights and Obligations to Future Generations for the Environment." https://scholarship.law.georgetown.edu/cgi/viewcontent.cgi?article=2639&context=facpub. Accessed 28 Sep. 2022.
5. Edith Brown Weiss, In Fairness to Future Generations: International Law, Common Patrimony, and Intergenerational Equity.originally published in 1988.
6. Redgwell, C. (1999). Intergenerational Trusts and Environmental Protection. Juris.
7. "Our Common Future: Report of the World Commission on ...." https://sustainabledevelopment.un.org/content/documents/5987our-common-future.pdf. Accessed 28 Sep. 2022.
8. "Rural Litigation And Entitlement ... vs State Of U.P. & Ors on 12 ...." https://indiankanoon.org/doc/1949293/. Accessed 1 Oct. 2022.
9. "Kinkri Devi And Anr. vs State Of Himachal Pradesh And Ors. on 29 ...." https://indiankanoon.org/doc/837514/. Accessed 1 Oct. 2022.
10. "Vellore Citizens Welfare Forum vs Union Of India & Ors on 28 .." https://indiankanoon.org/doc/1934103/. Accessed 1 Oct.2022.

Bibliography
❖ Statutes
- Environment Protection Act, 1986
- Indian Constitution Act, 1950

❖ International Conventions
- Ramsar Convention.
- Stockholm Convention
- CITES
- Convention on Biological Diversity (CBD)
- Bonn Convention
- Vienna Convention
- Montreal Protocol
- Kyoto Protocol

❖ Repots
- Stockholm conference report
- Brundtland Commission Report (Our Common future)
- The State of India's Environment 1982—A Citizen's Report
- The State of India's Environment, 1984–5: The Second Citizens' Report, by Anil Agarwal & Sunita Narain. Published by the Centre for Science and Environment,

❖ Articles
- Dalhousie Law Journal, Volume 30, Issue 1, Article 3 4-1-2007 Revisiting the Doctrine of Intergenerational Equity in Global Environmental Governance

❖ Website sources
- Manupatra
- Hein Online
- Inflibnet

Assistant Professor,
Patna Law College,
Patna University, Patna

www.ingramcontent.com/pod-product-compliance
Lightning Source LLC
Chambersburg PA
CBHW050252010526
44107CB00003B/285